AF564549

THE FUNGI AS PATHOGENIC AND BENEFICIAL MICROBES

NIPA GENX ELECTRONIC RESOURCES & SOLUTIONS P. LTD.
New Delhi-110 034

About the Author

Prof. S. G. Borkar, is an alumni of Indian Agricultural Research Institute, New Delhi for his Master and Doctorate degree in Plant Pathology (1977-1983), a French government scholar for post-doctorate at INRA, Angers, France in 1984, and had his Doctorate of Science (D.Sc) in Plant Pathology in 1999 from International University, Washington, USA.

As an academician and scientist of international repute, he has 130 research publications published over a period of 40 years in 17 foreign and in 39 Indian research journals besides 5 university publications and 2 book Chapters. Presented 29 scientific as well as Lead papers in national and international symposiums; published 8 books including USA publications, 6 patents, developed 4 wheat varieties, 2 documentary films, established 2 laboratories, 2 museums, mentored 30 students for their masters and doctorate degrees and received 21 national and International Awards for his contribution including a plant pathogenic klebsiella pneumoniae strain after his name by NCBI, USA, as Klebsiella pneumoniae strain Borkar. His Biographical note is included in Asia Pacific WHO'S WHO (vol.1, 1998) and Twentieth Century Admirable Achievers Distinguished WHO'S WHO (1999).

He served in 3 agricultural universities in 2 states of India in various academic, scientific and administrative positions including the post of Dean, Post-Graduate Institute. Lead the wheat research in peninsular India as zonal-co-ordinator at the national level for around 5 years and lead the university in plant pathology as a Head of the department for 10 years. He also rendered his services as Designated Inspection Authority of Plant Quarantine for Western Maharashtra, for the government of India, for ten years on the position of Head of Department of Plant Pathology at Mahatma Phule Agriculture University, Rahuri, Maharashtra state. After his superannuation from the university services of Mahatma Phule Agriculture University, Rahuri, in 2018, founded his own research laboratory" Endeavour Scientific Agriculture' as Startup (www.allagrisolutions.com) at Nashik in Maharashtra state, India, to serve the Indian farmers.

THE FUNGI AS PATHOGENIC AND BENEFICIAL MICROBES

Suresh G. Borkar
Former Professor and Head
Department of Plant Pathology
Mahatma Phule Krishi Vidyapeeth
Rahuri, Maharashtra, India

NIPA GENX ELECTRONIC RESOURCES & SOLUTIONS P. LTD.
New Delhi-110 034

NIPA GENX ELECTRONIC RESOURCES & SOLUTIONS P. LTD.

101,103, Vikas Surya Plaza, CU Block
L.S.C.Market, Pitam Pura, New Delhi-110 034
Ph : +91 11 27341616, 27341717, 27341718
E-mail:newindiapublishingagency@gmail.com
www: www.nipabooks.com
For customer assistance, please contact
Phone: + 91-11-27 34 17 17 Fax: + 91-11- 27 34 16 16
E-Mail: feedbacks@nipabooks.com

ISBN: 978-93-91383-83-1

Composed and Designed by NIPA.

Acknowledgement

I am thankful to my students Drs. Kalindee Shinde, Nivedita Kadam and Ajayasree T S who has helped me during the preparation of this manuscript. I am grateful to all my colleague professors and scientist across the world whose fungal photographs are used from public domain. In the absence of availability of their names in the public domain, the courtesy for their names in the photographs is not mentioned, however these are of those individual scientist who are anonymous for me.

My thanks are to all my professors of mycology and plant pathology in India and abroad whose inspirations serve as a souce to write such important title. My gratitude is to H.H. Sir Jaggu, H.H. Sir Fakira, H.H. Sir Deolu of Gopiwada and to my parents Sir Govinda and Sumitra, my wife Dr. Sandhya and son Antriksh for sparing me for this endeavour.

Prof. Suresh Borkar

Preface

Fungi are important microbes whose colonies can be seen with naked eyes on some of the infected hosts while their structures can only be visualized microscopically. These are present everywhere, in the atmospheric air, in water, water bodies and water currents, in soil and soil crust, on plant surfaces, vegetables and fruits, food grains and food product, on decaying plant material, on bodies of human, animals, birds, in marine environment and on marine life, on cotton clothes and leather articals in damp weather, on house ceilings in damp places and even in railways compartment ceilings. These are important to know beacause these are invasive to plant and plant products, human, animals, birds, marine life to cause diseases in them and sometimes deadly on their hosts. These differ from host to host and the diseases they causes.

Nevertheless, some other fungal species are beneficial to human as these are edible like mushrooms, used in manufacturing of enzymes, organic acids, in cheese industry, as colouring agents, in manufacturing of antibiotics and pharmaceuticals, as biological control agents, as biofertilizers and mineralizers, as decomposer of organic waste etc. These decomposers are also known as scavengers on the earth as they keep the earth and the environment free from garbages and its stinct.

In general, all over the world the public expect those who study the subject mycology, is unaware of the basic knowledge of fungi and their importance in their life. The book "The Fungi, as Pathogenic and Beneficial Microbe" deals with this basic knowledge which is essential of everyone including the students in schools and universities.

The content of the book is divided into three parts viz. An introduction to fungi, fungi harmful to living being and fungi useful to mankind and ecosystem and these parts are divided into twelve chapters with useful information on concern fungi, its invasive effect on concern host or beneficial nature with beautiful pictures.

It is a must collection in book rack of every home, libraries, and book shelves of various establishments to enrich the knowledge about this important microbe and the role they play in human life.

Prof. Suresh Borkar

Contents

Section III: Useful Fungi

Section I
An Introduction to Fungi

1

The Fungi and its Life

Fungi are very primitive eukaryotic microbe on the earth which are evolved with the earth and its atmosphere. These may be the ancestor of the plants in the process of evolution (Borkar, 2001). The group of ancestral fungi is thought to be represented by the present day chytridiomycota, although the Microsporidia may be an equally ancient sister group. The first major steps in the evolution of higher fungi were the loss of the chytrid flagellum and the development of branching, aseptate fungal filament, which occurred as terrestrial fungi diverged from water molds 600 million to 800 million years ago. The easily recognizable mushroom fungi probably diversified 130 million to 200 million years ago, soon after flowering plants became an important part of the flora and well before the age of dinosaurs. A relatively recent evolutionary radiation perhaps 60 million to 80 million years ago, of anaerobic chytridiomycota occurred as grasses and grazing mammals became abundant, the chytrid fungi serve as symbionts within the rumen of such animals, thereby enabling the grazing mammals to digest grasses.

These are omnipresent i.e can be found in soil, atmosphere, rocks, water streams, sea, air and surfaces of dead and living human beings, animals, birds, aquatic life, plants, food material and play an important role in the environmental equilibrium and ecosystem.

The world would have been garbage pill without the fungus which play their role as scavengers of these garbage and litter. Similarly, the world would have been with several diseases of human being without the antibiotics which the fungi gave us.

1.1. What is fungi

These are the macro and microorganisms. Macro in the sense that one can observed their presence with the naked eyes on the surface where it grows, while micro in the sense that their structures and spores are observed and detected only under microscope.

One can routinely observe their presence in most of the material kept in damp weather say for example bread, damaged fruits and vegetable and food grain, clothes, leather and several other articals. These form a white floppy or coloured growth on the surfaces on which they grow (Fig.1).

Fig. 1: Fungal colonies and structures of fungus

The fungal structure consists of a filamentous thread like structure called mycelium and several mycelia inter-woven together to form a colony or floppy structure is called thallus. The fungus mycelium may be divided in to individual cells having eukaryotic nucleus or without separate cells with suspended nucleus. This condition of the mycelium is called ceonocytic mycelium.

1.2. How do they multiply

Asexual forms of reproduction is kown in fungus where the spore bearing structure called conidiophores or sporophores arise from the mycelial mat and bears the spores or the spore bearing structure, the sporangium having the fungal spores (Fig.2).

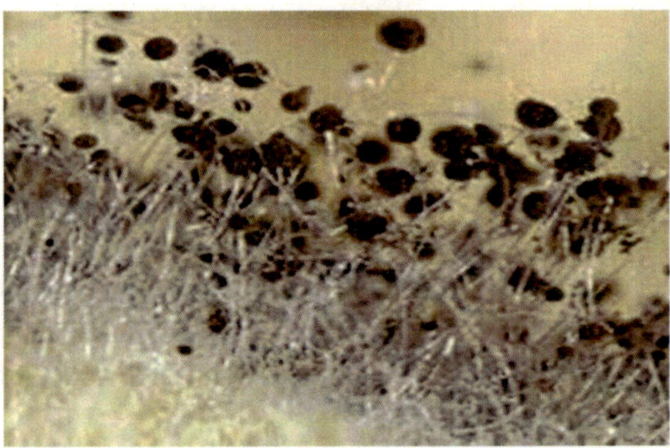

Fig. 2: Fungi bearing sporangium having fungal spores

Sometimes the mycelium itself divide into individual cells to act as new spore. The fungus spores are sometimes so delicate that these are wither in the absence of atmospheric humidity, and under dry weather or hot temperatures. In such fungal species, these spores are formed in the asexual fruiting bodies i.e. acervulus.

Thus fungus spores or reproductive units are formed when the fungus mycelium ages or when atmospheric conditions are not favourable for fungal growth, multiplication and survival. As the mycelium grows and become aged they forms their reproductive structures in the form of male and female organs (also termed as gamates; sperm and egg or oogonium as female sex organ and antheridium as male sex organ) which when comes in contact (different terms are used for this sexual mating like planogametic copulation, gametangial copulation, gametangial contact, spermatogamy and somatogamy etc (Fig.3), through copulation and fertilization, forms the sexual spore or fruiting bodies containing the sexual spores/seeds which upon germination form the true to type fungus colonies again.

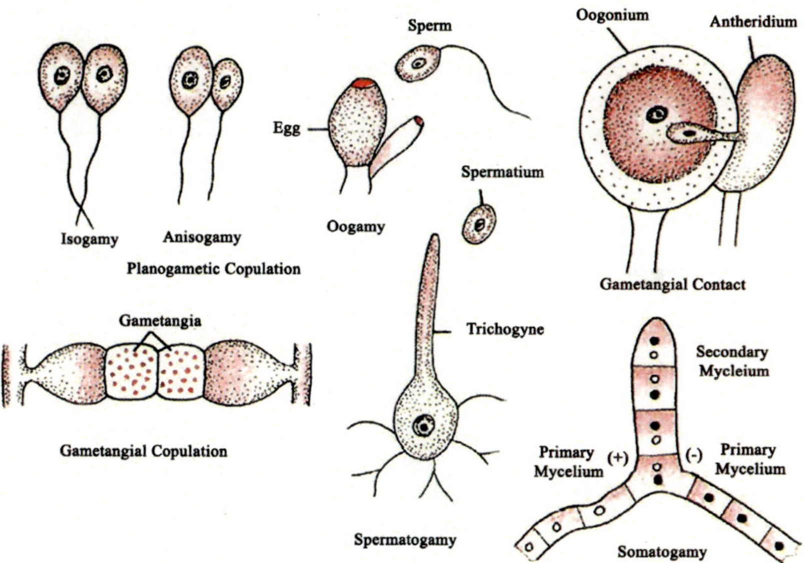

Fig. 3: Reproductive structures and sexual reproduction in fungi

1.3. How fungi obtain their nutrition

Like the human beings the fungus also requires carbohydrates, proteins and other essential mineral and organic acids for their growth, survival and reproduction. The fungus derives their food requirements from the host on which they grow. They absorb these food material from their host through absorptive mechanism.

In some fungal species a special kind of structure known as haustoria is produced from the growing surface mycelium which enters the host epidermal /cutical layer and derive their food by absorbing the sugar, amino acids and mineral molecules from host cellular content. Some species produce certain kinds of tissue macerating enzymes like cellulase (which degrade the cellulose molecules); pectinase (which degrade the pectate molecules) or proteinase (which degrade the protein molecules), so that these can be absorb by the growing fungus. Some fungus species produces toxins to kills their host cells and then from these killed cells they derive their nutrition by absorbing it. The fungus when grow on their pray, produced the structure called as appresorium, due to which the fungus body cling to its host surface and derive their nutrition. This nutrition is required for their growth and metabolic activities.

The fungus acquires the food either from living host material or dead decaying material and grow on them to derive their food. The fungal species which requires living host cells for their growth and multiplication are known as parasitic fungus species. These are unable to grow on artificial food media; while the fungal species which can grow on dead decaying material are called as saprophytes and these can be grown on artificial food media. Some fungal species are known to grow on both living host cells and in artificial food media are known as facultative parasites. Thus different types of fungal species have different types of food requirement.

1.4. The life cycle of fungi

The fungi produce asexual as well as sexual spores during their life cycle (Fig.4) which depend on the environmental condition. Under the favourable environmental condition for fungal growth, generally the asexual types of spores are produced. These are produced directly on the conidiophores or sporophores arising from the mycelium depending on the fungal species. Under unfavourable environment the asexual spores are also produced in the asexual fruiting bodies. The sexual spores are produced in the fruiting bodies formed due to interaction of sex organs of the fungi. These spores are the primary unit in the life cycle of the fungi. When the fungal spore come in contact with the suitable medium or host, it germinates by forming germ tube which grows into the mycelium. The mycelium in contact with the host cell form appresorium which cling to the host and send appendages known as haustoria into the host cell to absorb the nutrition from the host in case of ectoparasitic fungi. In endoparasitic fungi, the germ tube enters the host cell and develop into mycelium, absorb the nutrient to make them diseased. The mycelium produces the conidiophores and conidia/oidia as spore. The conidia disperse by wind or rain flashes to reach the new infection site to cause infection.

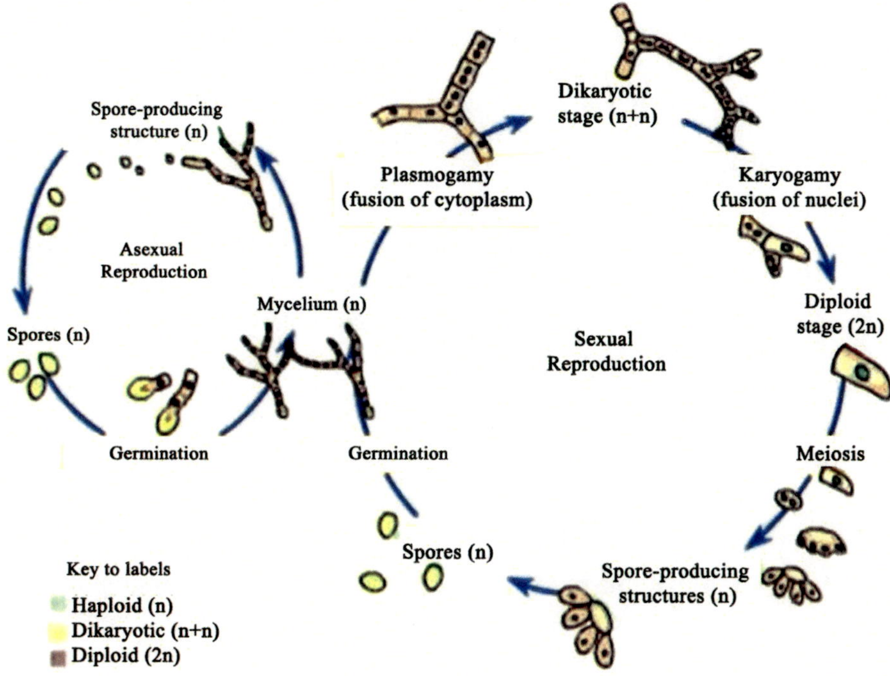

Fig. 4: Life cycle of fungi

Under unfavourable climate for fungal growth or as the fungus aged, it produced the asexual or sexual fruiting bodies depending on the fungal species. These contain spores which liberate from the fruiting bodies in the congenial condition and disperse by wind or rain flashes to reach the new infection site to cause infection.

In some fungal species the spores are very delicate and succumb to weathering condition while in others the spores are tough and form resting stage which resist to weathering condition. In saprophytic fungi the spores or mycelium lives on their substrate and multiplies in favourable weather condition.

1.5. Catagorization and classification of fungi

The fungus kingdom encompasses an enormous diversity of taxa with varied ecologies, life cycle strategies, and morphologies ranging from unicellular aquatic chytrids to large mushrooms. However, little is known of the true biodiversity of kingdom fungi, which has been estimated at 2.2 million to 3.8 million species. Of these, only about 120,000 have been described, with over 8,000 species known to be detrimental to plants and at least 300 that can be pathogenic to humans.

The true fungi, which make up the monophyletic clade is included in the kingdom fungi. The phylogenetic classification of kingdom fungi divides it into 7 phyla, 10 subphyla, 35 classes, 12 subclasses and 129 orders (Ainsworth & Bisby's Dictionary of Fungi 10th edition (2008). These seven major phyla are established according to their mode of sexual reproduction and using molecular data. Rapid advances in molecular biology and the sequencing of 18S r RNA (a part of RNA) continue to show new and different relationship between the various categories of fungi.

The seven true phyla of fungi are the Chytridiomycota (Chytrids), Blastocladiomycota, Neocallimastigomycota, Microsporidia, Glomeromycota, Ascomycota and Basidiomycota (the latter two being combined in the subkingdom Dikarya). These phyla are further subdivided into subphyla, classes, orders, families, genus and species on the basis of certain characteristics present in the fungus. The important characteristic of each phyla/phylum and their subdivision are as under:

1. Phylum: Chytridiomycota: The fungi of this phyla (Fig.5) are mainly aquatic, some are parasitic or saprotrophic; unicellular or filamentous; with chitin and glucan cell wall; primarily asexual reproduction by motile spores (zoospores): mycelia contain 2 classes **i.e Chytridiomycetes** and **Monoblepharidomycetes. The Chytridiomycetes** Contains 3 orders viz. Chytridiales, Rhizophydiales and Spizellomycetales while **Monoblepharidomycetes** Contain 1 order i.e Monoblepharidales

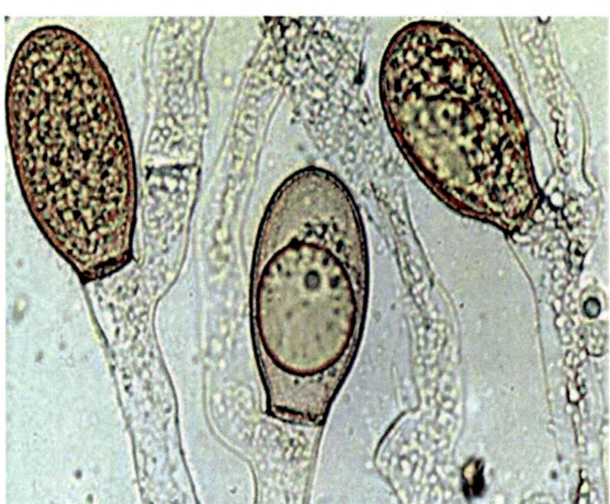

Fig. 5: Fungi of phylum Chytridiomycota

2. Phylum: Neocallimastigomycota: The fungi of this phyla (Fig.6) are found in the digestive tracts of herbivores; anaerobic; zoospores with one or more posterior flagella; lacks mitochondria but contains hydrogenosomes (hydrogen producing membrane bound organelles that generate energy in the form of ATP). The phyla contain 1 class i.e **Neocallimastigomycetes** which contain 1 order i.e Neocallimastigales

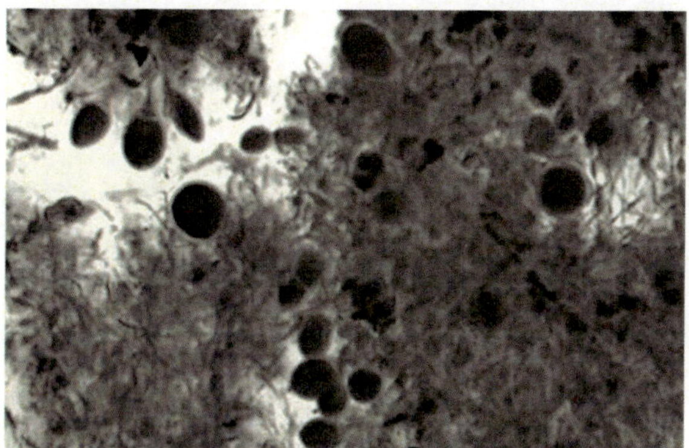

Fig. 6: Fungi of phylum Neocallimastigomycota

3. Phylum: Blastocladiomycota: The fungus of this phyla (Fig.7) are parasitic on plants and animals, some are saprotrophic, aquatic and terrestrial; flagellated; alternates between haploid and diploid generations (zygotic meiosis). It Contain 1 class i.e **Blastocladiomycetes** which contain 1 order i. e. Blastocladiales

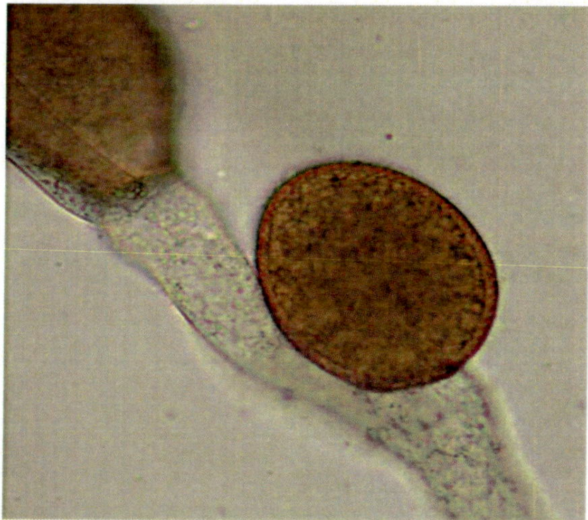

Fig. 7: Fungi of phyla Blastocladiomycota

4. Phylum: Microsporidia: The fungi of this phyla (Fig.8) are parasitic on animals and protists; unicellular, highly reduced mitochondria. Phylum not subdivided due to lack of well defined phylogenetic relationships within the group.

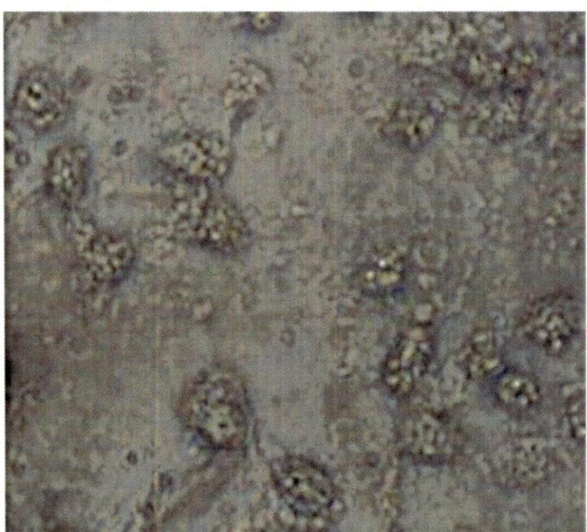

Fig. 8: Fungi of phyla Microsporidia

5. Phylum: Glomeromycota: The fungi of this phyla (Fig.9) forms obligate, mutualistic, or symbiotic relationships in which hyphae penetrate into the cells of roots of plants and trees (arbuscular mycorrhizal associations); coenocytic hyphae; reproduces asexually; cell walls composed primarily of chitin. It Contains 3 class i.e. **Archaeosporomycetes, Glomeromycetes and Paraglomeromycetes.**

The **Archaeosporomycetes** Contain 1 order i.e. Archaeosporales; the **Glomeromycetes** Contains 3 orders viz. Diversisporales, Gigasporales and Glomerales while **Paraglomeromycetes** Contains 1 order i.e Paraglomerales

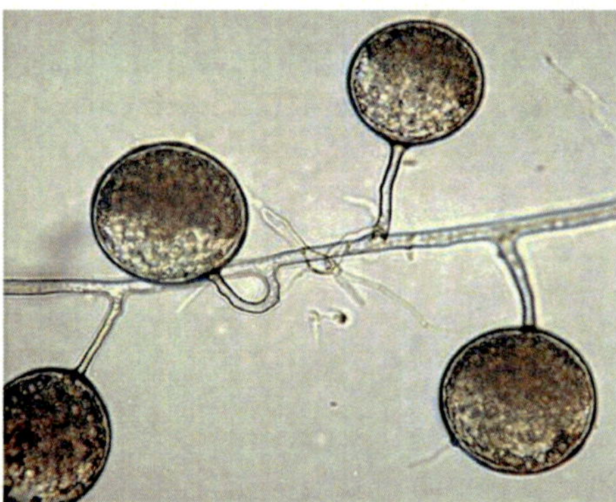

Fig. 9: Fungi of phyla Glomeromycota

Subphylum- Mucoromycotina: The fungi of this subphylum (Fig.10) are parasitic, saprotrophic, or ectomycorrhizal (forms mutual symbiotic association with plant); asexual or sexual reproduction; branched mycelium; It contains 3 orders that represent the traditional **zygomycota.** These orders are viz. **Mucorales (pin molds), Endogonales and Mortierellales.**

Fig. 10: Fungi of sub-phylum Mucoromycotina

Subphylum- Entomophthoromycotina: The fungi of this subphylum (Fig.11) are pathogenic, saprotrophic, or parasitic; coenocytic or septate mycelium; rhizoids formed by some species; conidiophore branched or unbranched; conidia forcibly discharged. It Contain 1 order i.e. **Entomophthorales**

Fig. 11: Fungi of subphylum Entomophthoromycotina

Subphylum- Zoopagomycotina: The fungi of this subphylum (Fig.12) are endoparasitic (lives in the body) or ectoparasitic (lives on the body) on nematodes, protozoa, and fungi; thallus branched or unbranched; asexual and sexual reproduction. It Contain 1 order i.e. **Zoopagales.**

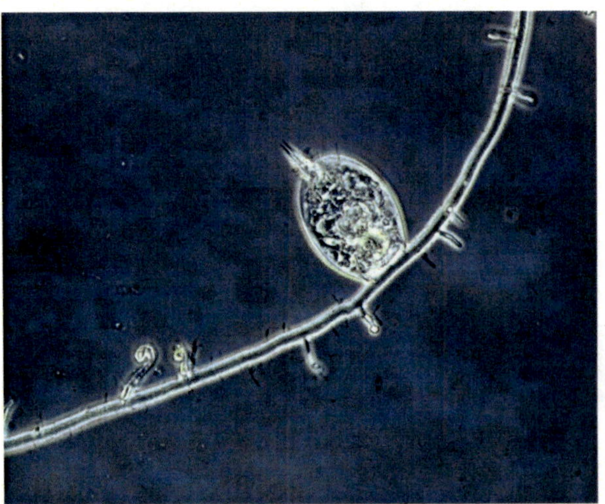

Fig. 12: Fungi of subphylum Zoopagomycotina

Subphylum- Kickxellomycotina: The fungi of this subphylum (Fig.13) are saprotrophic, or may be parasitic on fungi, can form symbiotic associations; thallus forms from holdfast on other fungi; mycelium branched or unbranched; asexual and sexual reproduction. It Contains 4 orders viz. **Kickxellales, Dimargaritales, Harpellales and Asellariales**

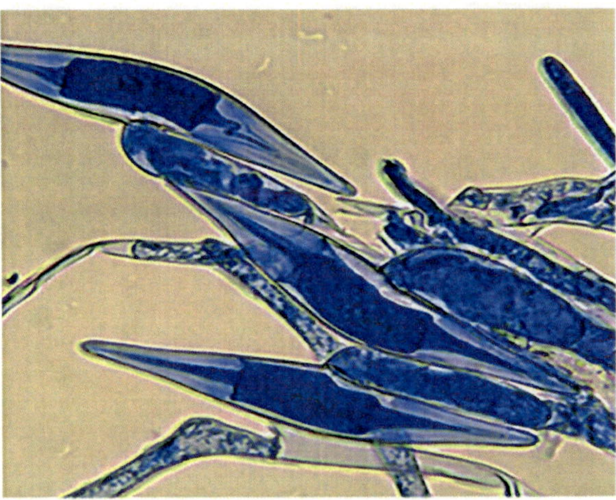

Fig.13. Fungi of subphylum Kickxellomycotina

6. Phylum: Ascomycota (Sac fungi) - The fungi of this phylum are symbiotic with algae to form lichens. Some are parasitic or saprotrophic on plants, animals, or humans; some are unicellular, but most are filamentous, hyphae septate with 1, rarely more, perforation in the septa, cells uninucleate or multinucleate, asexual reproduction by fission, budding or fragmentation or by conidia that are usually produced on sporiferous (spore-producing) hyphae, the conidiophores, which are borne loosely on somatic (main-body) hyphae or variously assembled in asexual fruiting bodies; sexual reproduction by various means resulting in the production of meiospores (ascospores) formed by free cell formation in saclike structures (asci),which are produced naked or, more typically, are assembled in characteristic open or closed bodies (ascocarps also called ascomata); Ascomycota include some cup fungi, saddle fungi, and truffles; this phylum is sometimes included in the subkingdom Dikarya with its sister group Basidiomycota

Subphylum: Taphrinomycotina- The fungi of this subphylum (Fig.14) are pathogenic on some plants, unicellular or filamentous, asci produced on the plant surface; ascocarp absent. It Contains 4 classes viz. **Taphrinomycetes, Neolectomycetes, Pneumocystidomycetes and Schizosaccharomycetes.** The Taphrinomycetes contain 1 order i.e. Taphrinales; the Neolectomycetes contain 1 order i.e. Neolectales, the Pneumocystidomycetes contain 1 order i.e. Pneumocystidales and Schizosaccharomycetes contain 1 order i.e. Schizosaccharomycetales (fission yeasts)

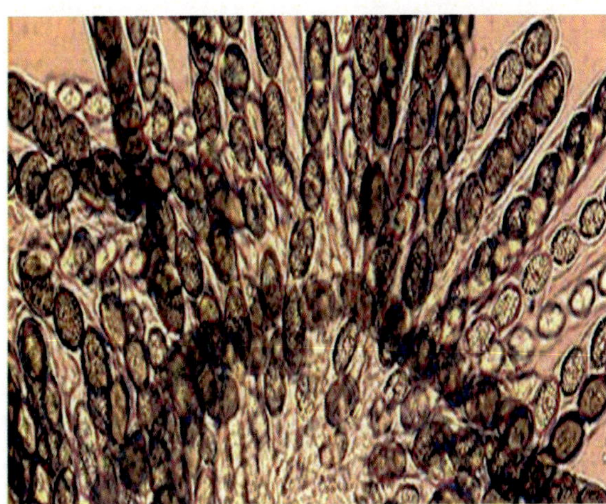

Fig. 14: Fungi of subphylum Taphrinomycotina

Subphylum: Saccharomycotina (True yeast)- The fungi of this subphylum (Fig.15) are saprotrophic on plants and animals, including humans, occasionally pathogenic in plants and humans; unicellular; found in short chains; asexual reproduction by budding or fission; contains common yeasts that are relevant to industry (e.g. baking and brewing) and that cause common infections in humans. It contains 1 class i.e. **Saccharomycetes** which contains 1 order i.e. **Saccharomycetales**

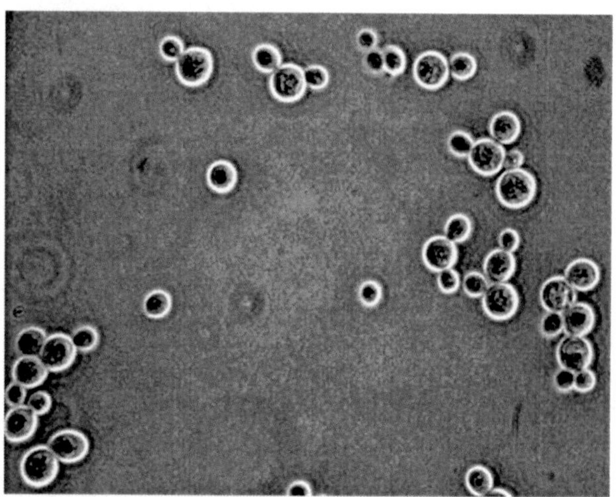

Fig. 15: Fungi of subphylum Saccharomycotina

Subphylum: Pezizomycotina- The fungi of this subphylum (Fig.16) are symbiotic with algae to form lichen; contains all ascomycetes able to produce ascomata; many form ascocarps, although some have lost the ability to undergo meiosis and cannot produce asci (formerly Deuteromycota). It contains 10 classes viz **Arthoniomycetes, Dothideomycetes, Eurotiomycetes, Laboulbeniomycetes, Lecanoromycetes, Leotiomyceytes, Lichinomycetes, Orbiliomycetes, Pezizomycetes, Sordariomycetes**

The **Arthoniomycetes** Contains 1 order i.e. **Arthoniales**, the **Dothideomycetes** contains 10 orders viz. **Capnodiales, Dothideales, Hysteriales, Jahnulales, Myriangiales, Pleosporales, Botryosphaeriales, Microthyriales, Patellariales, Trypetheliales**. The **Eurotiomycetes** contains 7 orders viz. **Chaetothyriales, Pyrenulales, Verrucariales, Coryneliales, Eurotiales, Onygenales and Mycocaliciales**. The **Laboulbeniomycetes** contains 2 orders viz. **Laboulbeniales and Pyxidiophorales**. The **Lecanoromycetes** Contains 10 orders viz. **Acarosporales, Lecanorales, Peltigerales, Teloschistales, Agyriales, Baeomycetales, Ostropales, Umbilicariales, Pertusariales, Candelariales** . The **Leotiomyceytes** contains 5 orders viz. **Cyttariales, Erysiphales (Powdery mildews), Helotiales, Rhytismatales and Thelebolales.** The **Lichinomycetes** Contains 1 order i.e. **Lichinales.**

The **Orbiliomycetes** contains 1 order i.e. **Orbiliales**. The **Pezizomycetes** contains 1 order i.e. **Pezizales**. The **Sordariomycetes** contains 19 orders viz. **Coronophorales, Hypocreales, Melanosporales, Microascales, Boliniales, Calosphaeriales, Chaetosphaeriales, Coniochaetales, Diaporthales, Ophiostomatales, Sordariales, Xylariales, Lulworthiales, Meliolales, Phyllachorales, Trichosphaeriales,**

Fig. 16: Fungi of subphylum Pezizomycotina

7. Phylum: Basidiomycota- The fungi of this phylum are parasitic or saprotrophic on plants or insects; filamentous; hyphae septate, with septa typically inflated (dolipore) and centrally perforated; mycelium are of two types: primary consisting of uninucleate cells. Succeeded by secondary consisting of dikaryotic cells, often bearing bridgelike clamp connections over the septa; asexual reproduction by fragmentation, oidia (thin-walled, free, hyphal cells behaving as spores), or conidia; sexual reproduction by fusion of hyphae (somatogamy), fusion of an oidium with a hypha (oidization), or fusion of a spermatium (a non motile male structure that empties its contents into a recepties of a receptive female structure during plasmogamy) with a specialized receptive hypha (spermatization), resulting in dikaryotic hyphae that eventually give rise to basidia, either singly on the hyphae or in variously shaped basidiocarps (also called basidiomata); meiospores (basidiospores) borne on basidia; in the rust and smuts, the dikaryotic hyphae produce teleutospores (thick walled resting spores), which are a part of the basidial apparatus; this is a large phylum of fungi containing the rust, smuts, jelly fungi, club fungi, coral and shelf fungi, mushrooms, puffballs, stinkhorns and bird's-nest fungi; sometimes included in the subkingdom Dikarya with its sister group, Ascomycota.

1. **Subphylum: Pucciniomycotina-** The fungi of this subphylum (Fig.17) are pathogens of land plants; includes the rusts. It contains 8 classes viz. Pucciniomycetes, Cystobasidiomycetes, Agaricostilbomycetes, Microbotryomycetes, Atractiellomycetes, Classiculomycetes, Mixiomycetes and Cryptomycocolacomycetes.

The Pucciniomycetes contains 5 orders viz. **Septobasidiales, Pachnocybales, Helicobasidiales, Platygloeales and Pucciniales**. The **Cystobasidiomycetes** contains 3 orders viz. **Cystobasidiales, Erythrobasidiales** and **Naohideales**. The **Agaricostilbomycetes** contains 2 orders viz. **Agaricostibales** and **Spiculogloeales**. The **Microbotryomycetes** contains 4 orders viz. **Heterogastridiales, Microbotryales, Leucosporidiales** and **Sporidiales**. The **Atractiellomycetes** contains 1 order i.e. **Atractiellales**. The **Classiculomycetes** contains 1 order i.e. **Classiculales**. The **Mixiomycetes** contains 1 order **i.e. Mixiales. The Cryptomycocolacomycetes** contains 1 order **i.e Cryptomycocolacales**.

Fig. 17: Fungi of subphylum Pucciniomycotina

Subphylum: Ustilaginomycotina- The fungi (Fig.18) are parasitic on plants as dikaryotic hyphae; haploid phase is saprotrophic. It contains 2 classes i.e. **Ustilaginomycetes** and **Exobasidiomycetes**. The **Ustilaginomycetes** contains 2 orders viz. **Urocystales** and **Ustilaginales**. The **Exobasidiomycetes** contains 7 orders viz. **Doassansiales, Entylomatales, Exobasidiales, Georgefischeriales, Malasseziales, Microstromatales** and **Tilletiales**.

Fig.18: Fungi of subphylum Ustilaginomycotina

Subphylum: Agaricomycotina- The fungi of this subphylum (Fig.19) are parasitic or symbiotic on plants, animals, and other fungi, some are saprotrophic or mycorrhizal; basidia may be undivided or have transverse or longitudinal septa; dolipore (inflated) septa and septal pore cap (parenthesomes) present; includes mushrooms, bracket fungi, puffballs. It contain 3 classes viz. **Tremellomycetes, Dacrymycetes** and **Agaricomycetes**. The **Tremellomycetes** contains 3 orders viz. **Cystofilobasidiales, Filobasidiales** and **Tremellales**. The **Dacrymycetes** contain 1 order i.e. **Dacrymycetales**. The **Agaricomycetes** contains 17 orders viz. **Agaricales, Atheliales, Boletales, Geastrales, Gomphales, Hysterangiales, Phallales, Auriculariales, Cantharellales, Corticiales, Gloeophyllales, Hymenochaetales, Polyporales, Russulales, Sebacinales, Thelephorales, and Trechisporales,**

Fig. 19: Fungi of subphylum Agaricomycotina

The fungal kingdom recently classified into 7 phylums were earlier classified into 6 phyla. The detail characteristics of these 6 phylums are as under.

1. Chytridiomycota: The Chytrids

The only class in the phylum chytridiomycota (Fig.20) is the chytridiomycetes. The chytrids are the simplest and most primitive eumycota or true fungi. The evolutionary records show that the first recognizable chytrids appeared during the late pre-Cambrian period, more than 500 million years ago. Like all fungi, chytrids have chitin in theire cell wall, but one group of chytrids has both cellulose and chitin in the cell wall. Most chytrids are unicellular; a few form multicellular organism and hyphae, which has no septa between cells (coenocytic). They produce gametes and diploid zoospores that swim with the help of a single flagellum.

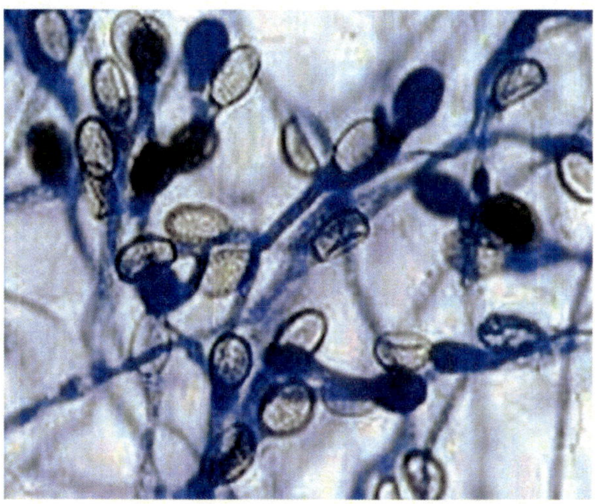

Fig. 20: Fungi of phylum Chytridiomycota

2. Zygomycota: The Conjugated fungi

The zygomycetes are a relatively small group of fungi belonging to the phylum zygomycota (Fig.21). They include the familiar bread mold *Rhizopus stolonifera*, which rapidly propagates on the surfaces of breads, fruits, and vegetables. Most species are saprobes, living on decaying organic material; a few are parasites, particularly of insects. Zygomycetes play a considerable commercial role. The metabolic products of other species of *Rhizopus* are intermediates in the synthesis of semi-synthetic steroid hormones.

Zygomycetes have a thallus of coenocytic hyphae in which the nuclei are haploid when the organism is in the vegetative stage. The fungi usually reproduce asexually by producing sporangiospores. The black tip of bread mold are the swollen sporangia packed with black spores. When spores land on a suitable substrate, they germinate and produce a new mycelium. Sexual reproduction starts when conditions become unfavourable. Two opposing mating strains (types + and type -) must be in close proximity for gametangia from the hyphae to be produced and fuse, leading to karyogamy. The developing diploid zygospores have thick coats that protect them from desiccation and other hazards.

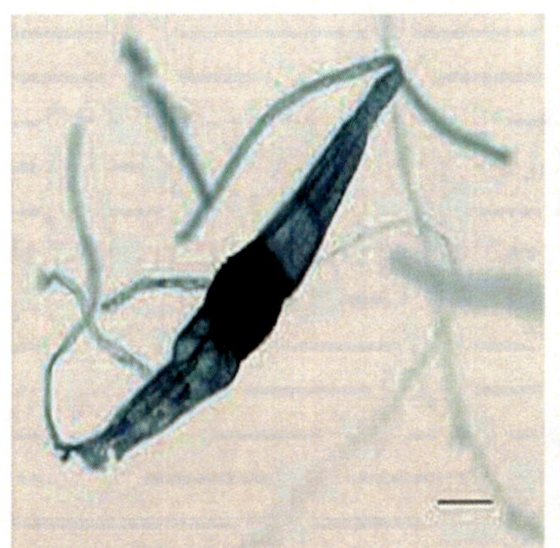

Fig. 21: Fungi of phylum Zygomycota

They may remain dormant until environmental conditions are favourable. When the zygospore germinates, it undergoes meiosis and produces haploid spores, which will in turn, grow into new organism. This form of sexual reproduction in fungi is called conjugation, giving rise to the name "conjugated fungi".

3. Ascomycota: The Sac fungi

The majority of known fungi belong to the phylum Ascomycota (Fig.22), which is characterized by the formation of an ascus, a sac–like structure that contains haploid ascospores. Many ascomycetes are of commercial importance. Some play a beneficial role, such as the yeast used in baking, brewing and wine fermentation, plus truffles and morels, which are held as gourmet delicacies. *Aspergillus oryzae* is used in the fermentation of rice to produce sake. Other ascomycetes parasitize plants and animals, including humans. For example, fungal pneumonia poses a significant threat to AIDS patients who have a compromised immune system. Ascomycetes not only infest and destroy crops directly; they also produce poisonous secondary metabolites that makes crops unfit for consumption. Filamentous ascomycetes produce hyphae divided by perforated septa, allowing streaming of cytoplasm from one cell to the other. Conidia and asci, which are used respectively for asexual and sexual reproduction, are usually separated from the vegetative hyphae by blocked (non-perforated) septa.

Asexual reproduction is frequent and involves the production of conidiophores that release haploid conidiospores. Sexual reproduction starts with the development of special hyphae from either one of the two mating strains. The male strain produces an antheridium and the female strain develops an ascogonium. At fertilization, the antheridium and the ascogonium combine in plasmogamy without nuclear fusion. Special ascogenous hyphae arise, in which pairs of nuclei migrate:one from the male strain and one from the female strain. In each ascus two or more haploid ascospores fuse their nuclei in karyogamy. During sexual reproduction, several of asci fill the fruiting body called the ascocarp. The diploid nucleus gives rise to haploid nuclei by meiosis. The ascospores are then released, germinate, and form hyphae that are disseminated in the environment and start new mycelia.

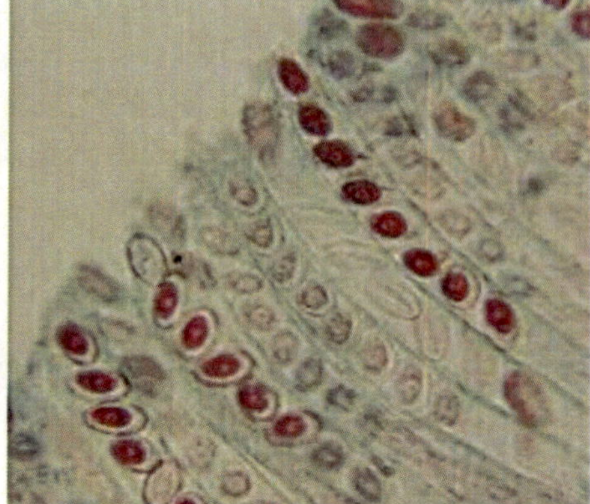

Fig. 22: Fungi of phylum Ascomycota

4. Basidiomycota: The Club fungi

The fungi in the phylum Basidiomycota (Fig.23) are easily recognizable under a light microscope by their club-shaped fruiting bodies called basidia (singular; basidium), which are the swollen terminal cell of the hypha. The basidia, which are the reproductive organs of these fungi, are often contained within the familiar mushrooms, commonly seen in the fields after rain, on the supermarket shelves, and growing on your lawn. These mushroom producing basidiomyces are sometimes referred to as "gill fungi" because of the presence of gill like structures on the underside of the cap. The gills are actually compacted hyphae on which the basidia are borne. This group also includes shelf fungus, which cling to the bark of trees like small shelves. In addition, the Basidiomycota includes smut and rust, which are important plant pathogens; toadstools, and shelf fungi stacked on tree trunk. Most edible fungi belong to phylum Basidiomycota; however, some Basidiomycota produce deadly toxins, for example *Cryptococcus neoformans* causes severe respiratory illness.

Fig. 23: Fungi of phylum Basidiomycota

The life cycle of basidiomycetes includes alternation of generation. Spores are generally produced through sexual reproduction, rather than asexual reproduction. The club shaped basidium caries spores called basidiospores. In the basidium, nuclei of two different mating strains fuse (Karyogamy), giving rise to a diploid zygote that then undergoes meiosis. The haploid nuclei migrate into basidiospore, which germinate and generate monokaryotic hyphae. The mycelium that results is called a primary mycelium. Mycelia of different mating strains can combine and produce a secondary mycelium that contains haploid nuclei of two different mating strains. This is the dikaryotic stage of the basidiomyces lifecycle and it is the dominant stage. Eventually, the secondary mycelium generates a basidiocarp, which is a fruiting body that protrudes from the ground- this is what we think of as a mushroom. The basidiocarp bears the developing basidia on the gills under its cap.

5. Asexual Ascomycota and Basidiomycota

Imperfect fungi- those that do not display a sexual phase-use to be classified in the form phylum Deuteromycota, a classification group no longer used in the present. While Deuteromycota use to be a classification group, recent molecular analysis has shown that the members classied in this group (Fig.24) belong to the Ascomycota or the Basidiomycota classification. Since they do not posses the sexual structures that are used to classify other fungi, they are less well described in comparison to other members. Most members live on land, with a few aquatic exceptions. They form visible mycelia with a fuzzy appearance and are commonly known as mold.

Fig. 24: Fungi of phylum Asexual Ascomycota and Basidiomycota

Reproduction of the fungi in this group is strictly asexual and occurs mostly by production of asexual conidiospores. Some hyphae may recombine and form heterokaryotic hyphae. Genetic recombination is known to take place between the different nuclei.

6. Glomeromycota

The Glomeromycota (Fig.25) is a newly established phylum which comprises about 230 species that all live inclose association with the roots of trees. Fossil records indicate that trees and their root symbionts share a long evolutionary history. It appears that all members of this family form arbuscular mycorrhizae: the hyphae interact with the root cells forming a mutually beneficial association where the plant supply the carbon source and energy in the form of carbohydrates to the fungi, and the fungus supplies essential minerals from the soil to the plants

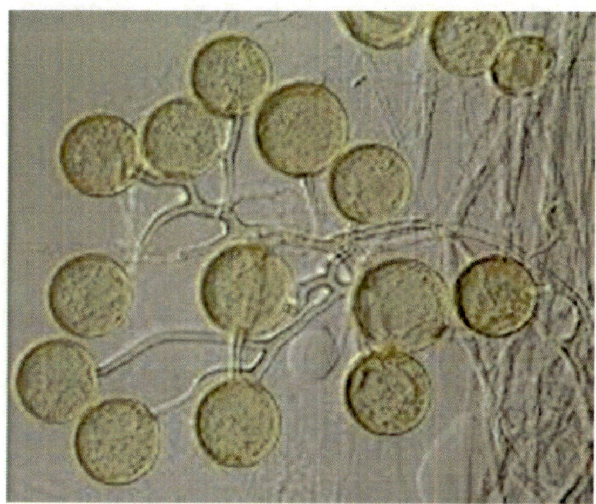

Fig. 25: Fungi of phylum Glomeromycota

The glomeromycetes do not reproduce sexually and do not survive without the presence of plant roots. Although they have coenocytic hyphae like the zygomycetes, they do not form zygospore. DNA analysis shows that all glomeromycetes probably descended from a common ancestor, making them a monophyletic lineage.

The fungal nomenclsture is binomial, with a generic and species name, for example *Aspergillus oryzae*, where *Aspergillus* is a generic name while *oryzae* is a species name. The species are grouped in genera, genera into families (suffix- aceae), families in orders (suffix-ales) and orders in classes (suffix-mycetes) and classes into phylum.

1.6. Impact of fungi on mankind and environment

Fungi are involved in a wide range of activities. Some are useful while some are harmful to mankind. Their useful activities vary from decomposition of natural waste material on earth, mineralization of rocks to form soil particles, enrichment of soil through solubilisation and mobilization of plant nutrients. Some fungi produce antibiotics to be used as human and animal medicines while others produce organic acids and petroleum products to be used in industries. Some fungi produce plant growth promoters and alkaloids. Some fungi are used as food material. Some fungi are parasites of crops insect pest and disease causing organisms thus works as bio-control agent to control of pest and diseases, maintaining the ecological balance in nature. The fungal fruiting bodies like mushrooms can be used as food while the fruiting bodies of bracket fungi like Ganoderma is used as medicine.

Most of the fungi survive on the waste material available in the natural ecosystem while some fungi require the live host for their growth, survival and metabolic activities. These live hosts include plants, human beings, animals, birds, fishes etc and causes various types of disease symptoms on them. These diseases are harmful to the respective host leading to their death and losses. Various parasitic fungi act as causal organisms and infect hundreds of species and varieties of plants of the economic value. There are thousands of species of plant pathogenic fungi that collectively are responsible for 70% of all known plant diseases. They also spoil fruits, vegetables and all kinds of foodstuff. They cause diseases in important crops, with enormous losses affecting the economies and survival of the mankind. Such known diseases are late blight of potato, black stem rust of wheat, brown spot of rice, etc

Fungal infections presently destroy at least 125 million tonnes of the top five food crops viz. rice, wheat, maize, potatoes and soybeans each year, which could otherwise be used to feed those who do not get enough to eat. The damage caused by fungi to rice, wheat and maize alone costs global agriculture $60 billion per year. The effects are disproportionately catastrophic for those in the developing world, where 1.4 billion people live on less than $1.25 per day, and rely most heavily on these low-cost foods. Diseases like rice blast, soybean rust, stem rust in wheat, corn smut in maize and late blight in potatoes affect more than just productivity; many have wide ranging socio-economic costs.

They induce diseases in animal and human also, e.g., mycosis caused by infestation of *Aspergillus, Cercospora* and *Cryptococcus*. Fungi are common habitant in the environment and some are even considered normal inhabitants of the skin, gastrointestinal tract and other mucous membrane surfaces. It also causes infectious diseases in Birds and Aquatic animal/fish also. Fungal infections in fish can cause damage to multiple body systems, such as the liver, kidney, and brain, and usually occur when the fish is in a weakened state, either due to injury or trauma. In animals, new fungal diseases increasingly threaten the existence of over 500 species of amphibian, as well as many endangered species of bees, sea turtles and corals. In the US alone, studies suggest the decline in bat populations caused by white nose syndrome fungus will lead to a dramatic rise in the insect crop-pests that the bats would otherwise eat, and a cost to agriculture of more than $3.7 billion per year.

Trees lost or damaged by fungi are equated to trees fail to absorb 230-580 megatonnes of atmospheric CO_2, equivalent to 0.07% of global atmospheric CO_2, an effect the scientists say is likely to be leading to an increase of the greenhouse effect.

2

The Habitat of Fungi

The fungi are present everywhere. They are in the soil and soil crust, in atmospheric air, in sea and pond water, rivers and water streams, and on the surfaces of almost all the material ,whether living or otherwise, on affected food stuff , vegetables and food grains, on body of human beings, animals, birds and fishes, on plant surfaces, on leathers and clothes in damp weather, on house walls and ceiling in damp conditions, on railway coaches and buses roofs having water leakage and dampness, in caves, on garbage and decomposing material etc. The fungal spores are present everywhere, however the fungal growth occurs where the congenial conditions for their growth is available.

2.1. In air

Air content several species of fungi (Fig.26) as spores of fungi are commonly dispersed in wind. The air we breathe contains spores of many different fungi. The spores of common air borne fungi have thick melanised wall. Melanin content in fungal spore wall is first barrier to UV radiation and therefore the fungal spores survive in adverse environments.

The common fungal species found in indoor and outdoor environment are several species of *Aspergillus* **particularly** *flavus, fumigatus, niger, versicolor, caespitosus, candidus, carneus, clavatus, ochraceus, restrictus, sydowii, terreus, unguis, ustus,* **and** *wentii.* **Other common fungal species found in the air are species of** *Acremonium, Alternaria, Aureobasidium, Candida, Cladosporium, Curvularia, Epicoccum, Fusarium, Geotrichum, Hyalodendron, Penicillium, Rhinocladiella, Rhodotorula, Trichoderma, Tritirachium, Beauveria, Bipolaris, Botryosporium, Botrytis, Calcarisporium, Chaetomium, Chrysonilia, Chrysosporium, Cunninghamella, Drechslera, Eupenicillium, Eurotium, Exophiala, Gliocladium, Gliomastix, Humicola, Memnoniella, Monocillium, Mucor, Nigrospora, Oedocephalum, Oidiodendron, Ostracaderma, Paecilomyces, Pestalotia, Peziza, Phialophora, Pithomyces, Rhizoctonia, Rhizopus, Scopulariopsis, Sporobolomyces, Sporothrix, Syncephalastrum, Torulomyces, Trichocladium, Trichothecium, Ulocladium, Verticillium, Wallemia and Zygosporium.*

Fig. 26: Spores of fungal species in air

Inhalation of fungal spores in most cases has no effect on human. The spores lodge on the moist surface of the lining of the airways and they are subsequently expelled in mucus. The remaining spores are neutralised by the immune responce. However, a few fungi evade the immune responce and cause respiratory disease viz.

A. Asthama

Asthama is a severe allergic response in the lungs to allergens. The cells lining bronchi and alveoli become inflamed, severely reducing the capacity to inhale. Among common allergens are the spores of some fungi like ***Alternaria, Cladosporium and Aspergillus.***

These fungi are common saprotroph and endophyte on many plants and sporulates profusely especially as the leaf senesces. Spores of the fungi are usually dispersed only a short distance in high densities. Once airborn, spores may disperse long distances in the air stream. These spore disperses are common during crop harvesting and handling and therefore in most of the places the cases of asthama are increased during the period of crop harvestings.

B. Farmer's Lung

The disease is commonly found in people who handle hay and compost. ***Aspergillus fumigatus*** can lodge and grow in the lungs where it may cause **Aspergillosis**. The fungus is a thermotolerant and thermostable saprotrophic commonly isolated from compost especially during high temperature stage of composting. Thus farmer's lung is a syndrome associated with composting and old plant material handling.

The other fungi which are involved in farmer's lung disease are ***Cryptococcus neoformans, Coccidiodes immitis*** *and* ***Histoplasma capsulatum.*** These are commonly found in soils or in birds dropping as in case of *C. neoformans*. The spores are distributed in wind. When the spores lodge in the lung of susceptible individuals, the development of pneumonia like symptoms is followed by systemic infection, possibly involving the central nervous system.

2.2. On decomposing material and waste

Decomposing material and waste are the habitat of several fungi, which help in the process of decomposition. These fungi produce several enzymes which disintegrate the cellular material and convert them into useful biomolecules. Several fungi are found in this habitat.

2.2.1. Epicoccum nigrum

Also found in **air** and **soil** (Fig.27).

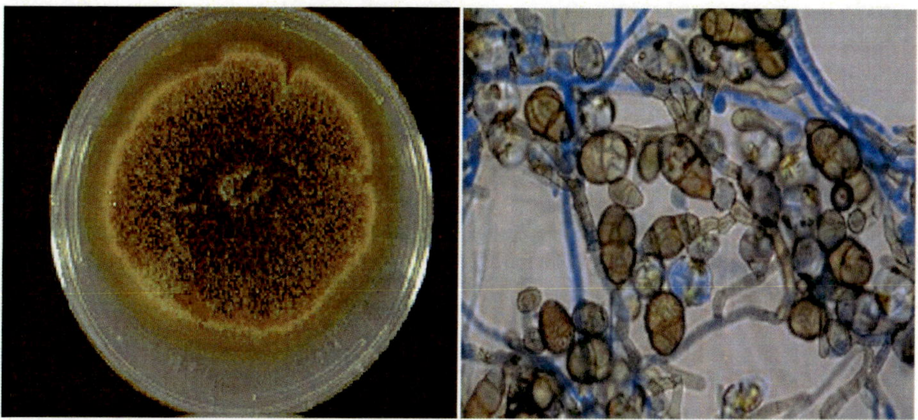

Fig. 27: Fungal colony and spores of Epicoccum nigrum

2.2.2. Coprinus comatus

Coprinus comatus (Fig.28), the **shaggy ink cap**, **lawyer's wig**, or **shaggy mane**, is a common fungus often seen growing **on lawns**, along **gravel roads** and **waste areas**.

Synonyms - *Agaricus cylindricus* Schaeff. (1774), *Agaricus comatus* O.F.Müll. (1780*), Agaricus vaillantii* J.F.Gmel. (1792).

Fig. 28: Fungal fruiting bodies of *Coprinus comatus*

Coprinus sterquilinus

Also called as Midden Inkcap (Fig.29) this mushroom has been found on **weathered horse dung, occasionally on rabbit dung, and on rotten plant debris**. An important identification feature is that this fungus is found on weathered dung rather than on the ground.

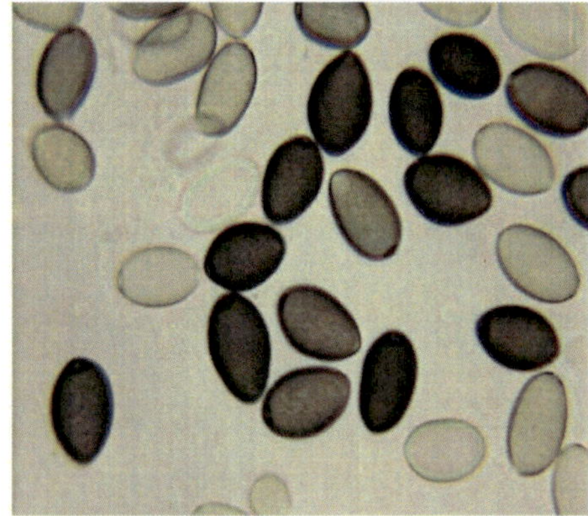

Fig. 29: Fruiting bodies and fungal spores of *Coprinus sterquilinus*

2.2.3. Leptosphaeria maculance

Natural habitat of *Leptosphaeria maculance* (Fig.30) includes **leaves and culms of a great many grasses, dead herbaceous stems,** and **driftwood.**

Synonyms

*Phyllosticta brassicae, Sphaeria maculans Sowerby (*1803).

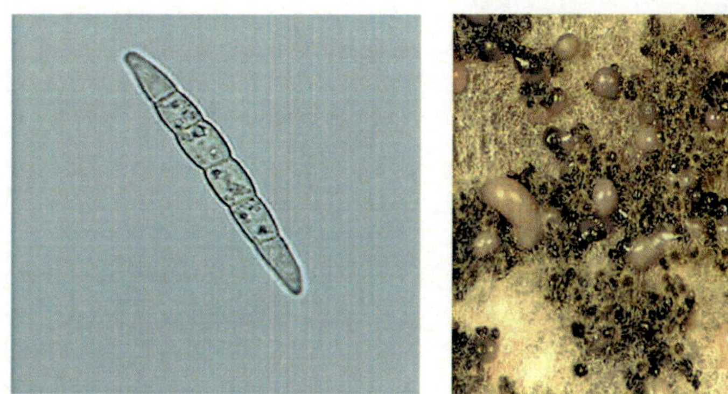

Fig. 30: Fungal growth and spores of Leptosphaeria maculance

2.2.4. Ascobolus immersus

Ascobolus immersus Pers. Fr. (Fig.31) is a cosmopilitan heterothallic ascomycota (Pezizales) living on **herbivore dung** (coprophilous).

Synonyms

Phaeopezia subgen. Crouaniella Sacc. (1884), Ascobolus subgen. Dasyobolus Sacc. (1889), Dasybolus Clem. & Shear (1931), Anserina Velen (1934), Seliniella Arx & E. Müll. (1955)

Fig. 31: Fungal structure of Ascobolus immersus

2.2.5. Pilobolus spp.

Pilobolus (Fig.32) is a genus of fungi that commonly grows on herbivore dung.

Synonyms

Hydrogera F.H. Wigg.ex Kuntze (1891), Pycnopodium Corda (1842)

Fig. 32: Fungal structure of *Pilobolus spp.*

2.2.6. Podospora fimiseda

Podospora species (Fig.33) is one of the most common *coprophilous ascomycetes* living **in dung** of many herbivores worldwide.

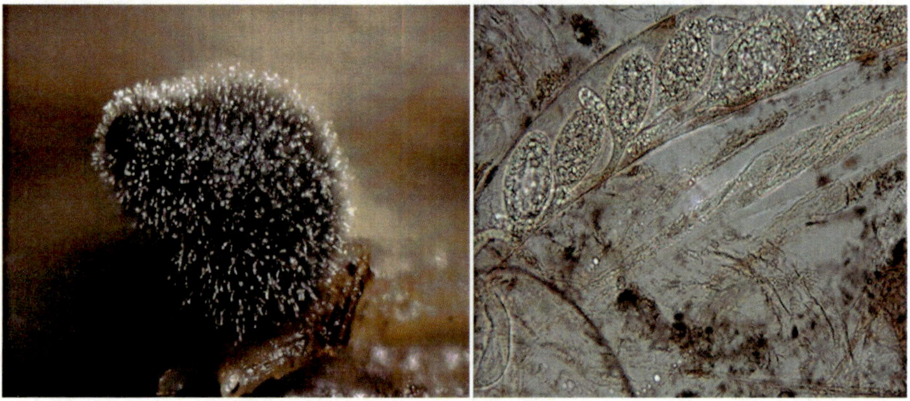

Fig. 33: Fungal structure and ascospores of *Podospora species*

2.2.7. Laccaria laccata

Laccaria laccata (Fig.34) is found in scattered troops in **wooded areas**, and on heathland often **in poor soil.**

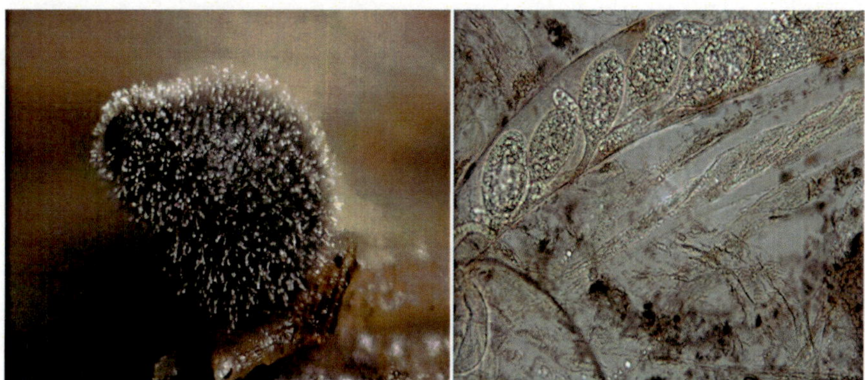

Fig. 34: Fungal fruiting bodies of *Laccaria laccata*

2.2.8. Chaetomium species

Chaetomium (Fig.35) is a dematiaceous, filamentous fungus belonging to a large genus of saprobic ascomycetes found in **plant debris, soil, air, dung, straw, paper, bird feathers, and seeds.**

Chaetomium globosum

Synonyms

Chaetomium kunzeanum Zopf, Chaetomium affine Corda, Chaetomium setosum Bainier, Chaetomium barbatum Traaen, Chaetomium subterraneum Swift & Povah, Chaetomium japonicum Saito & Okasaki.

Fig. 35: Fungal structure and spores of Chaetomium sp.

2.2.9. Paecilomyces variotii

Paecilomyces variotii (Fig.36) is a common environmental mold that is widespread **in composts, soils, food products, indoor air, wood, and carpet dust.**

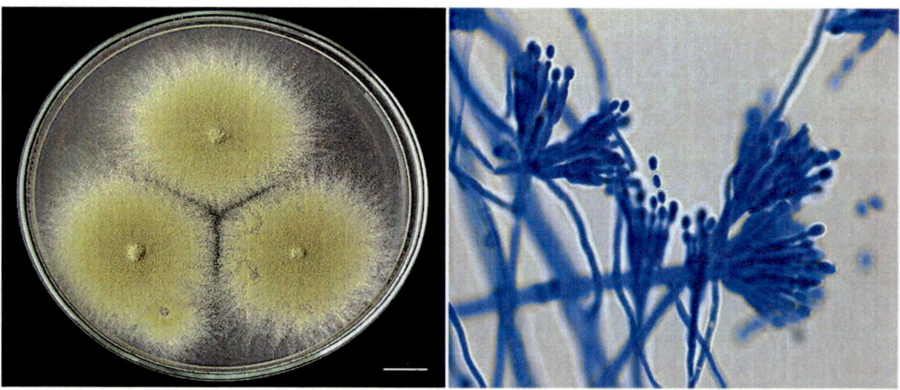

Fig. 36: Fungal colony and spore bearing structure of Paecilomyces variotii

2.3. On wood and wood logs

The habitat of several fungi are wood and wood logs.

2.3.1. Rhinocladiella species

Rhinocladiella (Fig.37) is a cosmopolitan fungus which can be found in **soil, herbaceous substrates,** and **decaying wood.**

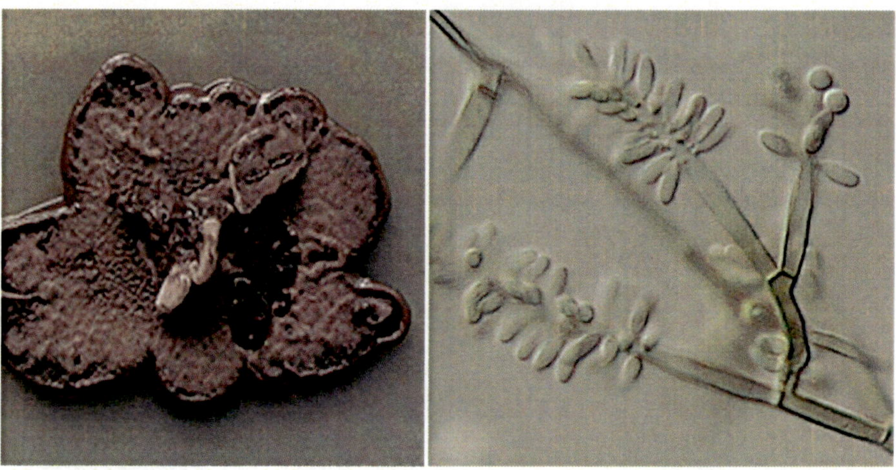

Fig. 37: Fungal and spore bearing structure of Rhinocladiella sp.

2.3.2. Ganoderma species

Ganoderma is a genus of polypore mushrooms that grow on **wood**.

Ganoderma lucidum

Ganoderma lucidum (Fig. 38) also known as Reishi or Lingzhi, typically grows **on hardwoods.**

Fig. 38: Fruiting bodies of Ganoderma lucidum

Ganoderma applanatum

Ganoderma applanatum (Fig.39), also known as the artist's bracket, artist's conk or bear bread.

Synonyms

Boletus applanatus, Elfvingia applanata, Fomes applanatum, Fomes vegetus, Ganoderme aplani, Ganoderma lipsiense, Polyporus applanatum, Polyporus vegetus.

Fig. 39: Fruiting bodies and spores *of Ganoderma applanatum*

Ganoderma multipileum

Ganoderma multipileum (Fig.40), commonly known as lingzhi or chizhi.

Fig. 40: Fruiting bodies of *Ganoderma multipileum*

Ganoderma tsugae

G. tsugae (Fig.41) tends to grow on conifers, especially hemlocks.

Fig. 41: Fruiting bodies of Ganoderma tsugae

2.3.3. Varicellaria rhodocarpa

Varicellaria rhodocarpa (Fig.42) grows and found on bark of **conifers.**

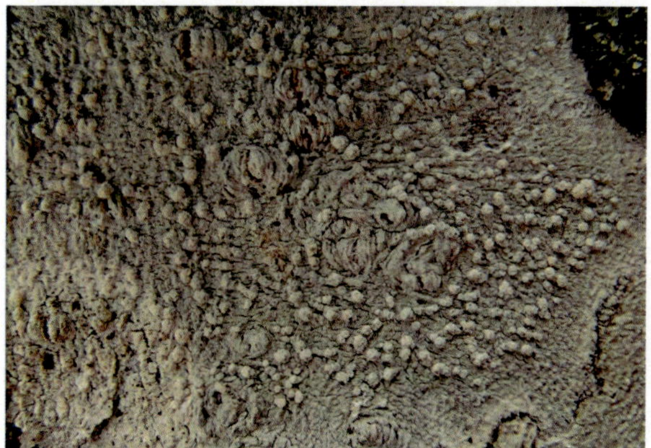

Fig. 42: Growth and fruiting bodies of Varicellaria rhodocarpa

2.3.4. Cordyceps subsessilis

Cordyceps subsessilis (Fig.43) has been found in **organic soils** or **decaying wood**. The host of *Cordyceps subsessilis* is a beetle of the family Coleoptera.

Fig. 43: Fruiting bodies of *Cordyceps subsessilis*

2.3.5. Mollisia species

It is a wood-rotting cup fungus most frequently found (Fig.44) on **dead hardwood**, but also occasionally **on conifers**.

Fig. 44: Fruiting bodies of Mollisia sp.

2.3.6. Daldinia concentrica

The inedible fungus ***Daldinia concentrica*** (Fig.45) is known by several common names, including **King Alfred's Cake**, **cramp balls**, and **coal fungus**. It lives on **dead and decaying wood**, especially on felled ash trees

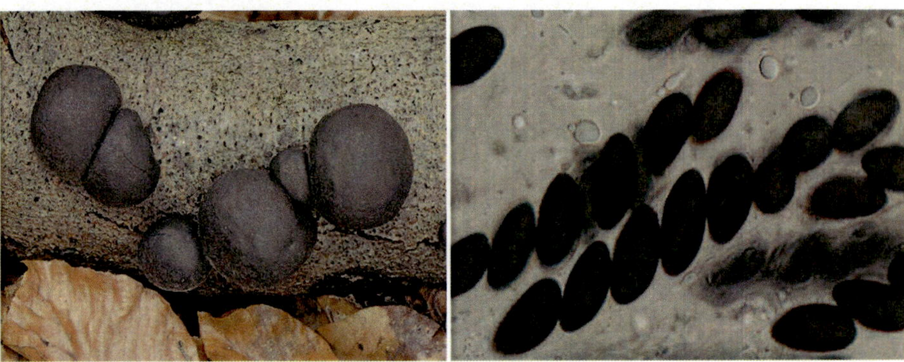

Fig. 45: Fruiting bodies of *Daldinia concentrica*

2.4. In soil and soil crust

Soil is an oligotrophic medium for the growth of fungi. Nutrients for the fungal growth are extremely limited for most of the time in barren unculted soils. The presence of living plant material or dead plant material dramatically changes the scenario of fungal population. Plants exude organic matter and they feed arbuscular mycorrhizal fungi. Other fungi survive on dead plant material.

The life styles adopted by fungi in soil is either ruderals, mycorrhizal or hyphal. Ruderals take advantage of the flushes of nutrients usually associated with rainfall. The water moving through soil carries with it the dissolved organic molecules flushed from plant surfaces, surface litter and dead microbes. The fungi respond immediately, and grow actively while soil remains moist. The fungi usually sporulate rapidly and exit through dry periods as asexual spores.

The fungi play an important role in soil, degrading complex sources of organic carbon, some of which may be organic pollutant. An increase in the deposition of aromatic carbon such as melanin in micro-aggregates may reverse losses of carbon associated with cultivation. Fungi that aggregate soil must attach to the soil particles and the hypaae must remain intact. Several groups of fungi appear to be involved with aggregating soil. AM fungi are the most important soil fungi for aggregation. The saprophytic fungi which survive on dead plant material help in decomposition of the waste and humus formation, while other soil fungi are disease causing agents on/in the root system of the plants.

The common generas of soil fungi include ***Acremonium, Absidia, Fusarium, Verticillium, Aspergillus, Penicillium, Chaetomium, Mucor, Alternaria, Mortierella, Rhizopus, Sclerotium, Pythium, Phytophthora, Syntrichium, Rhizoctonia, Ceratocystis** and **species of Glomus.***

2.4.1. Acremonium species

It is also known as *Cephalosporium* (Fig.46).

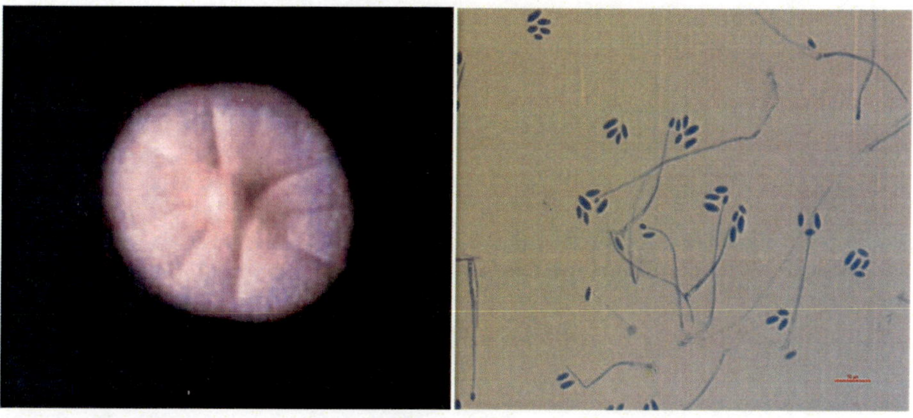

Fig. 46: Fungal growth and spores of Acremonium sp.

2.4.2. Fusarium oxysporum (Fig.47)

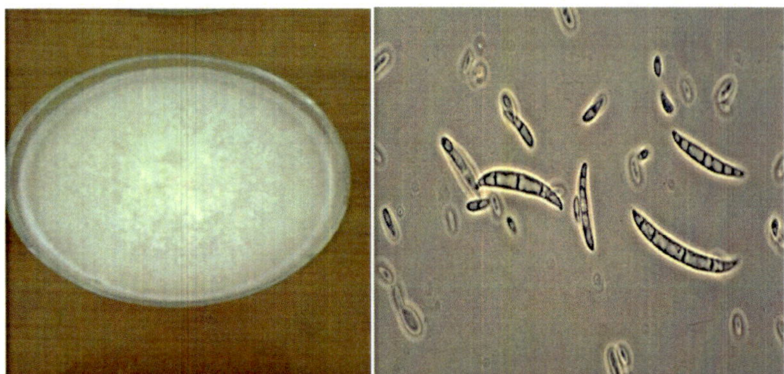

Fig. 47: Fungal growth and spores of *Fusarium oxysporum*

Fusarium solani and Fusarium moniliforme (Fig.48)

Fig. 48: Fungal growth of Fusarium solani and Fusarium moniliforme and spores of fusarium.

2.4.3. Penicillium bilaiae (Fig.49)

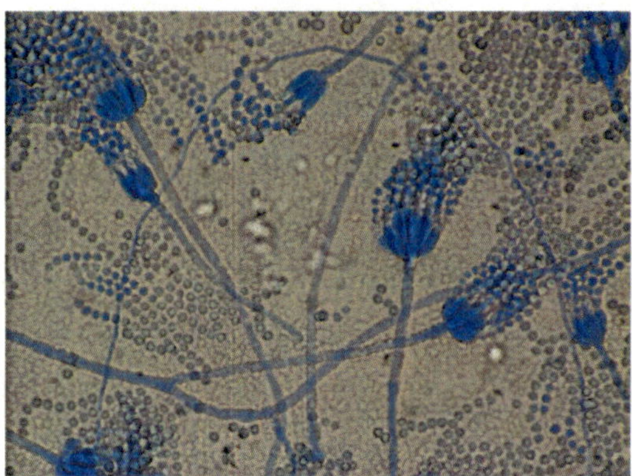

Fig. 49: Fungal structure and spores *of Penicillium bilaiae.*

2.4.4. Geotrichum spp

Geotrichum is a genus of fungi (Fig.50) found worldwide in **soil, water, air,** and **sewage.**

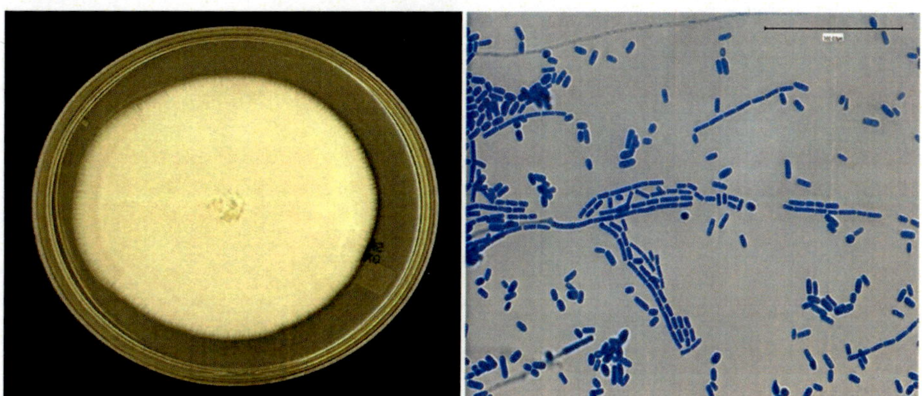

Fig. 50: Fungal colony and spores of *Geotrichum* spp.

2.4.5. Trichoderma species

Trichoderma is a genus of fungi that is present in all **soils**, where they are the most prevalent culturable fungi.

Trichoderma harzianum (Fig.51)

Synonyms

Sporotrichum narcissi Tochinai & Shimada, (1930), Trichoderma lignorum var. narcissi (Tochinai & Shimada) Pidopl. (1953), Trichoderma narcissi (Tochinai & Shimada) Tochinai & Shimada, (1931)

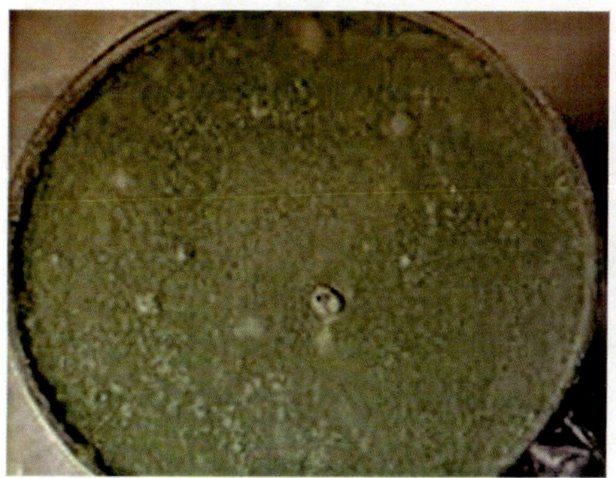

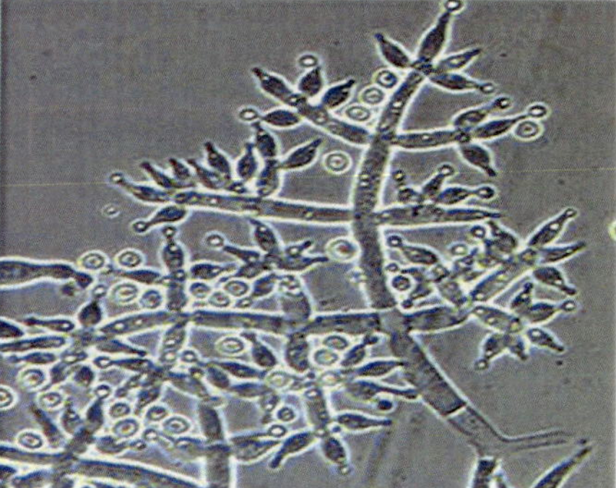

Fig. 51: Fungal colony and spores structures of *Trichoderma harzianum*

Trichoderma viride (Fig.52)

Synonyms

Hypocrea contorta (Schwein) Berk & M.A. Curtis, (1875), Hypocrea rufa (Pers.) Fr., Summa veg. Scand., (1849), Hypocrea rufa f. sterilis Rifai & J. Webster, (1966), Hypocrea rufa var. rufa (Pers.) Fr., Summa veg. Scand., (1849), Pyrenium lignorum Tode, (1790), Sphaeria contorta Schwein., (1832), Sphaeria rufa Pers., (1796).

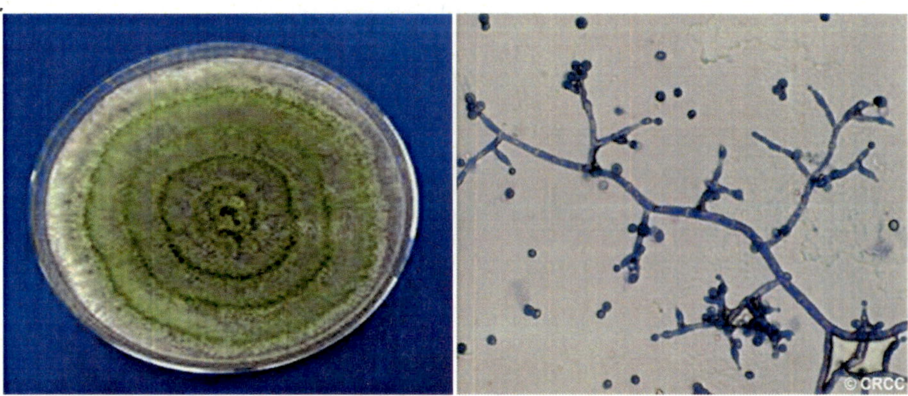

Fig. 52: Fungal colony and spores structures of *Trichoderma viride*

2.4.6. Chaetomium spp.

Chaetomium (Fig.53) is a genus of fungi found in **soil, air, cellulose containing material and plant debris.**

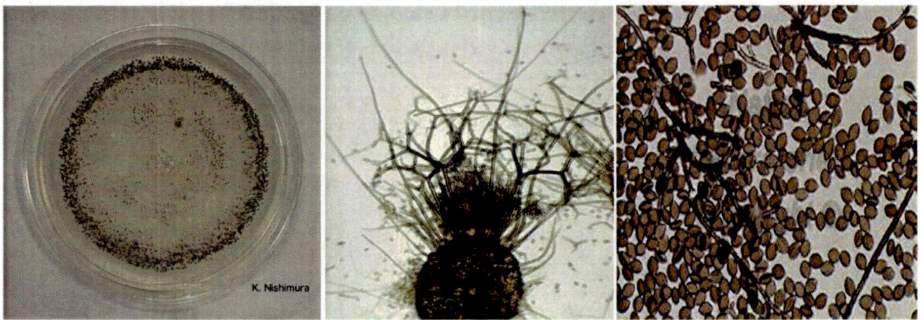

Fig. 53: Fungal colony, fruiting body and spores of *Chaetomium* spp.

2.4.7. Chromelosporium species

Chromelosporium species (Fig.54) are common **in soil** but are not well known.

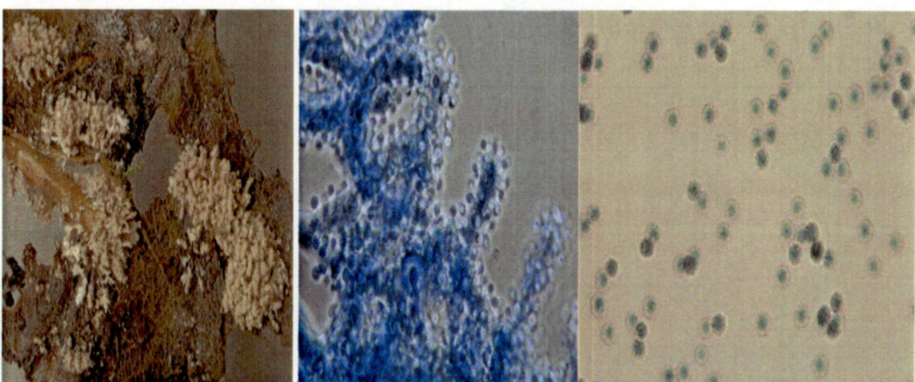

Fig. 54: Fungal structure and spores of Chromelosporium sp.

2.4.8. Ustilago Species (Fig.55)

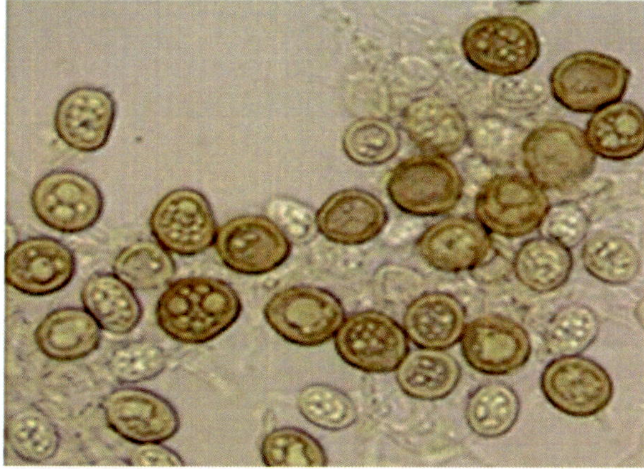

Fig. 55: Fungal spores of Ustilago spp.

2.5. In water and water bodies

Aquatic fungi include those fungi present in water, terrestrial fungi that release spores which are dispersed in water and the species that function entirely within water. Water is found in transient pools after rain, swamps, estuaries, permanent rivers and oceans. Fungi living in these environments require a wide array of characteristics to cope with the constraints. Conditions in fresh water are markedly different to those in the ocean. The open sea has quite different salinity to salterns where the water is slowly evaporating. Sometimes the water is extremely acidic or toxic. Fungi are found in all these environments.

A. Ingoldian fungi

These fungi are a characteristic group of hyphomycetes with specialised spores. In streams the major input of plant material comes from leaf fall. These material contain a range of endophytes of terrestrial origin and these endophytes are soon replaced by a characteristic group of hyphomycetes with specialised spores. Like the common terrestrial fungi, the Ingoldian fungi includes many species that appear to lack specificity to the host plant.

Fungi found in aquatic habitats have to gain access to organic materials to maintain themselves. Ingoldian (Fig.56) and other aquatic fungi appear to have evolved unique life histories. The Ingoldian fungi are essentially biphasic, with one life component dispersing the fungi over land, and the second taking advantage of the transient flushes of nutrients following leaf fall to water. Fungi have important roles in water. Apart from diminution of leaf litter, the aquatic chytrids appear to be important regulators of the picoplankton in water.

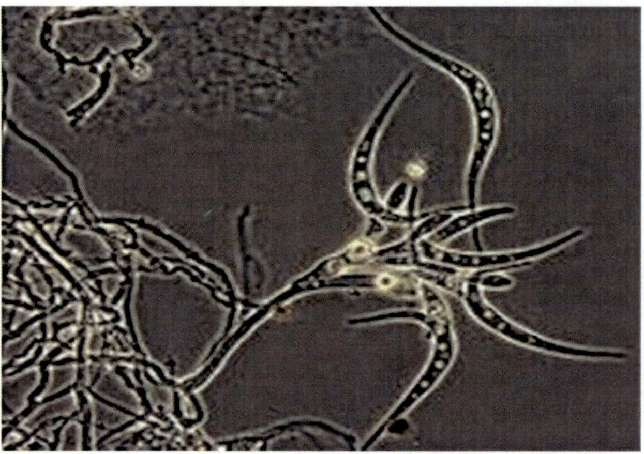

Fig. 56: Fungal spores of Ingoldian fungi.

2.5.1. Rhodotorula portillonensis sp. *nov*

Rhodotorula portillonensis sp. nov (Fig.57), *a basidiomycetous yeast is present in shallow-**water marine sediment** in Fildes Bay, King George Island, Antarctica.*

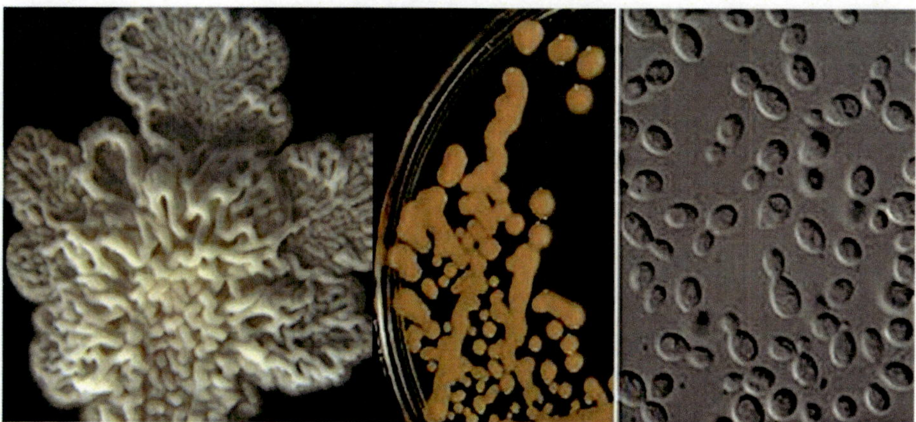

Fig. 57: Fungal colony characteristic and spores of *Rhodotorula portillonensis*

2.5.2. Leptosphaeria pelagica

Leptosphaeria pelagica (Fig 58) is having marine habitat.

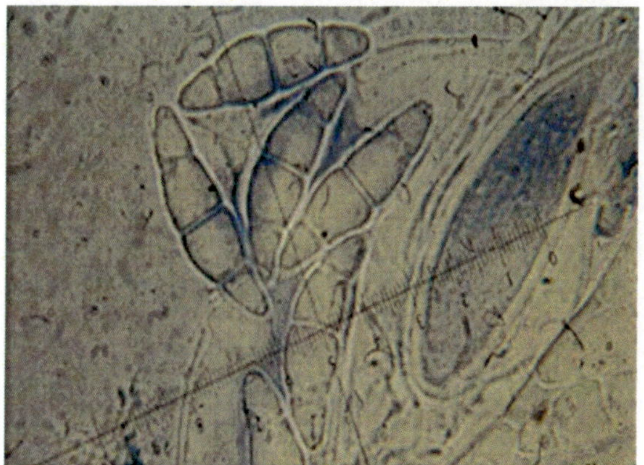

Fig. 58: Fungal spores of *Leptosphaeria pelagica*

2.5.3. Loramyces juncicola

The species *Loramyces juncicola* (Fig.59) was originally found on **growing dead, partly softened stems of *Juncus militaris* that were submerged in fresh water** to a depth of a few inches to as much as a few feet in a pond on Nashawena Island, off the Massachusetts coast.

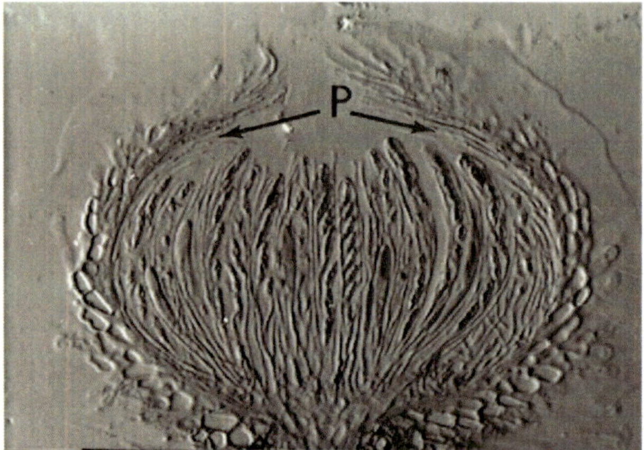

Fig. 59. Fungal fruiting body of *Loramyces juncicola with asci*

2.5.4. Annulatascus species

The genus (Fig.60) is characterized by taxa that are saprobic on submerged, decaying plant material in freshwater habitats.

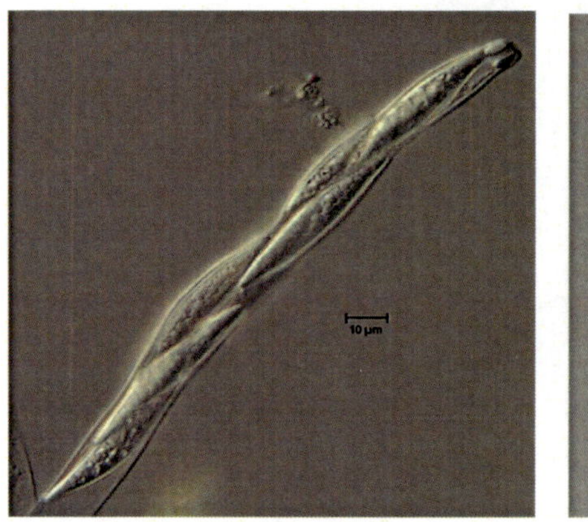

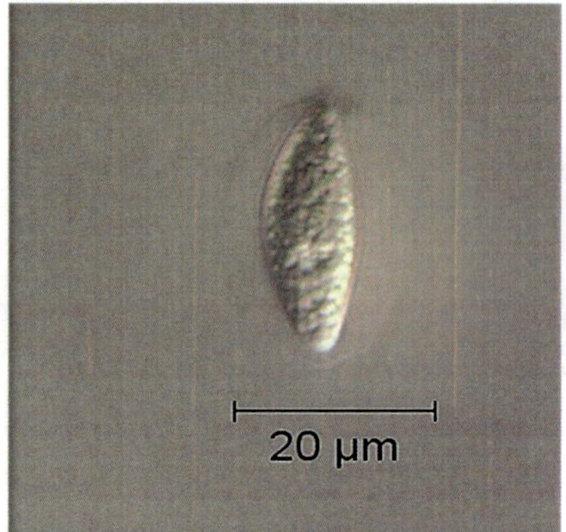

Fig. 60: Fungal ascospores of *Annulatascus species*

2.5.5. Proboscispora species

A freshwater ascomycete genus *Proboscispora* (Fig.61) found on **wood submerged in streams** in Australia and the Philippines.

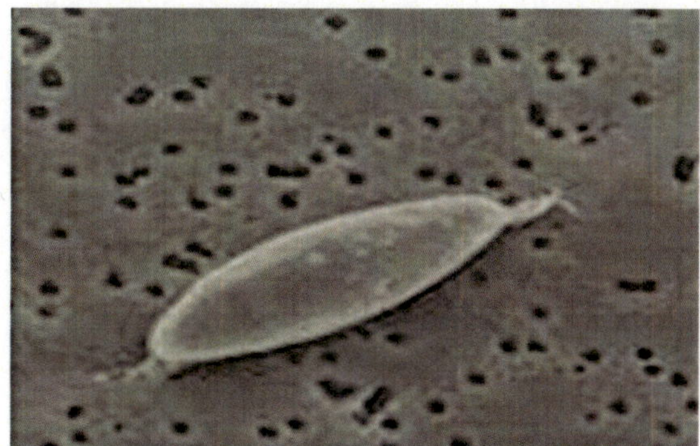

Fig. 61: Fungal ascospores of proboscispora sp.

2.5.6. Savoryella species

Savoryella (Fig.62) is one of the most commonly reported unitunicate ascomycetous genera from **submerged wood in rivers and streams and the marine environment.**

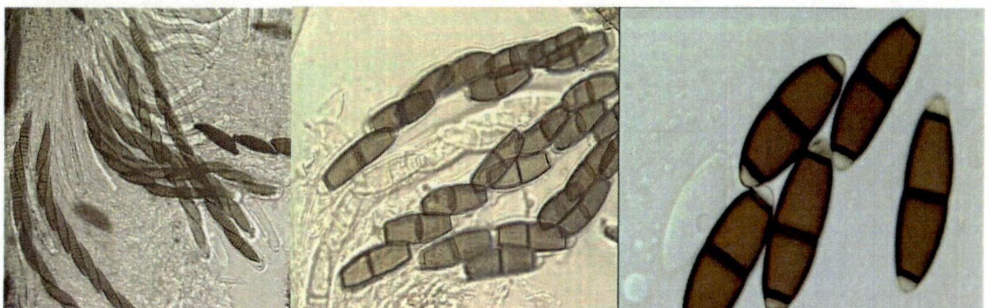

Fig. 62: Asci and ascospores of *Savoryella sp.*

2.5.7. Ascotaiwania

Ascotaiwania (Fig.63) is found **on driftwood in a freshwater stream and terrestrial palms.**

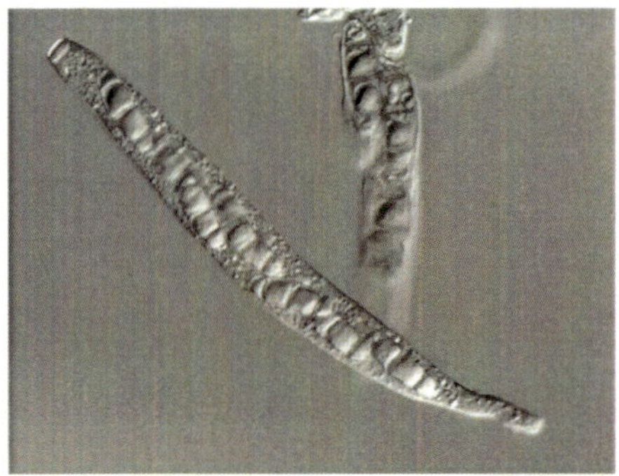

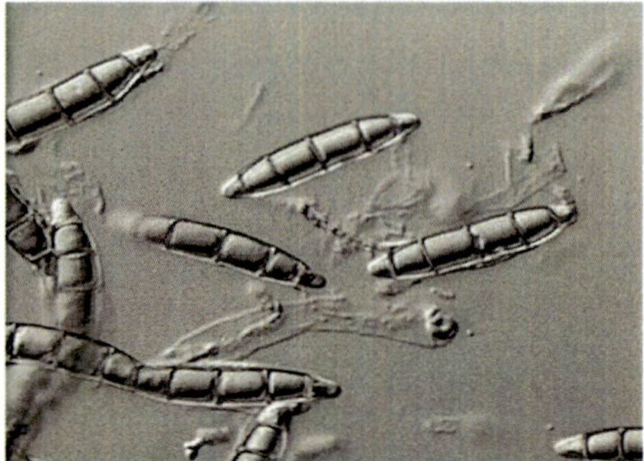

Fig. 63: Asci and ascospores of *Ascotaiwania* sp.

2.5.8. Aniptodera species

It is found in **marine** and **fresh water** habitat (Fig.64)

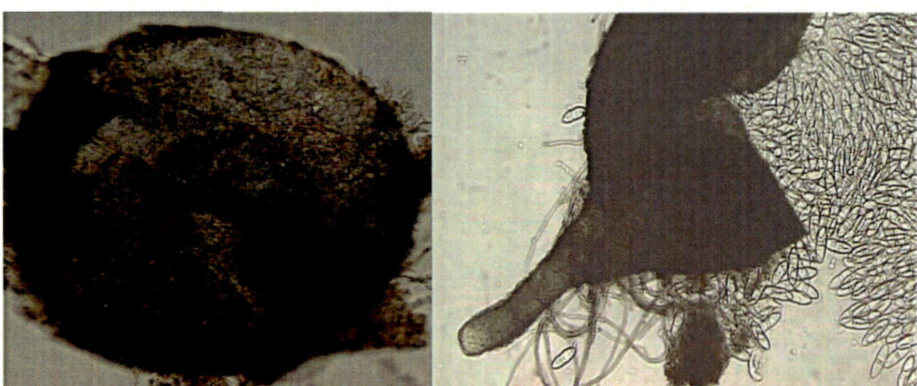

Fig. 64: Fruiting bodies and spores of Aniptodera sp.

2.5.9. Halotthia species

It is found in **marine habitat** (Fig.65).P

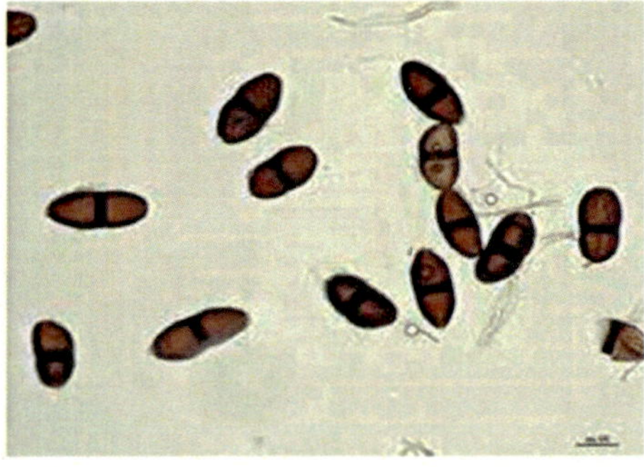

Fig. 65: Fruiting bodies and spores of *Halotthia* sp.

2.5.10. Leptosphaeria oraemaris

Leptosphaeria species (Fig. 66) is inhabiting **submerged wood in marine habitat.**

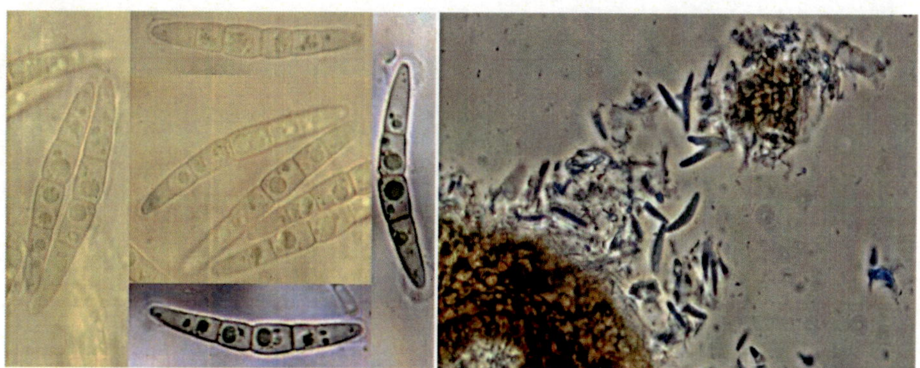

Fig.66: Fungal spores of *Leptosphaeria* species

2.5.11. Mycosphaerella species

It is found in **marine habitat** (Fig. 67).

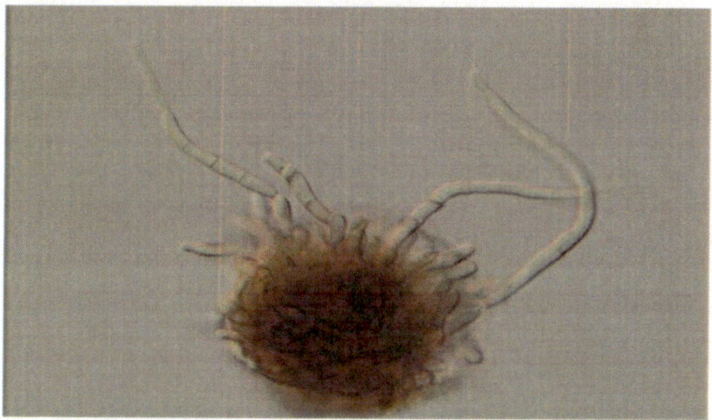

Fig.67: Fruiting bodies of *Mycosphaerella* species

2.5.12. Chaetosphaeria species (Fig.68)

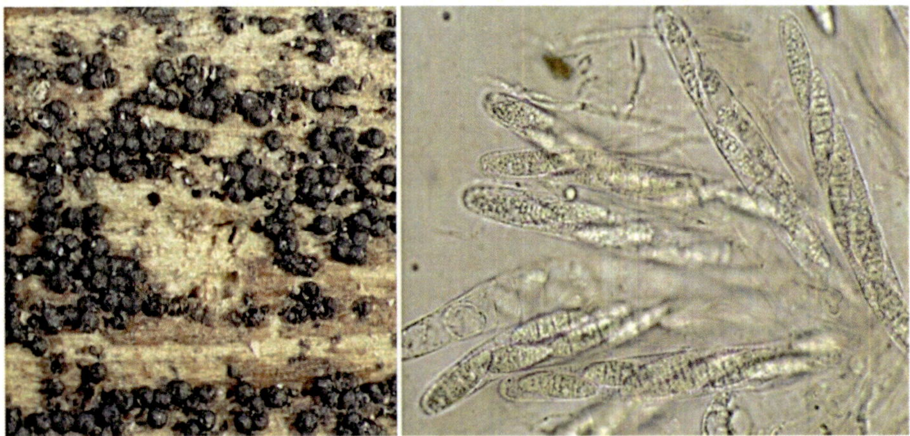

Fig. 68: Fruiting bodies and spores of *Chaetosphaeria* sp.

2.5.13. Hydronectria kriegeriana

A lichenized ascomycete (Fig.69) **on rocks in fresh water**.

Synonyms-*Kallichroma Tethys*

Fig. 69: Fruiting bodies of hydronectria kriegeriana

2.5.14. Kallichroma tethys

It is found **on dead mangrove wood** (Fig.70).

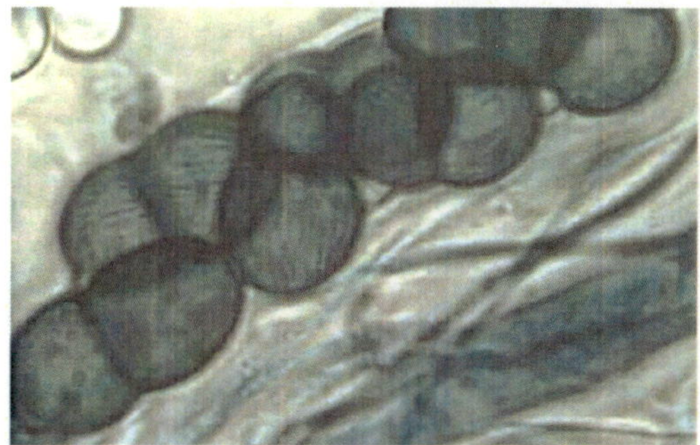

Fig. 70: Ascospores of Kallichroma tethys

2.5.15. Ceriosporopsis species

Ceriosporopsis species (Fig.71) is a primary **marine fungus**. It grows **on wood** aggressively and is suitable for biopulping of both soft and hardwoods.

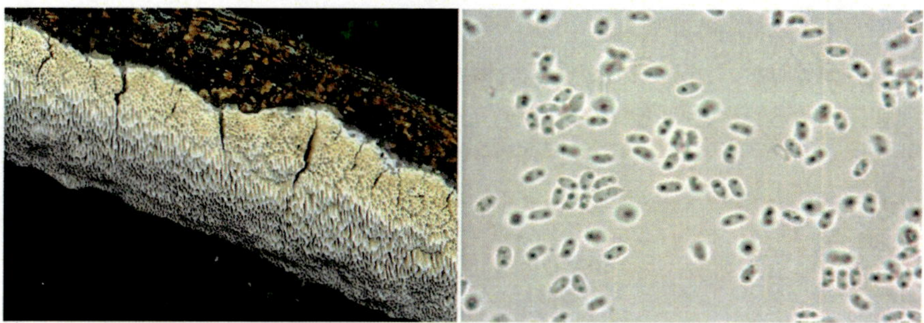

Fig. 71: Fungal growth on wood and spores of Ceriosporopsis species

2.5.16. Corollospora matrima-

It is also called as **marine sac fungus** (Fig.72).

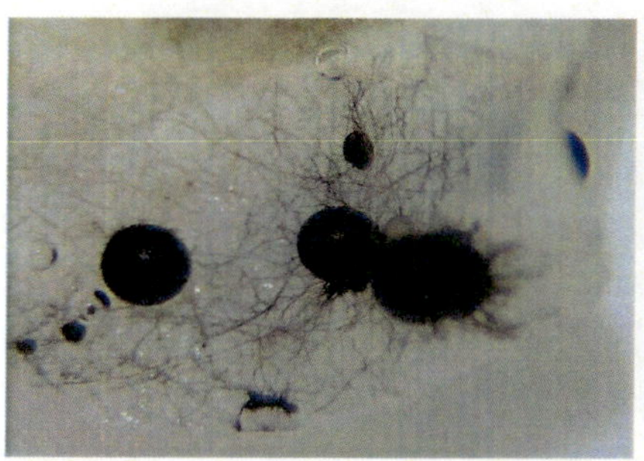

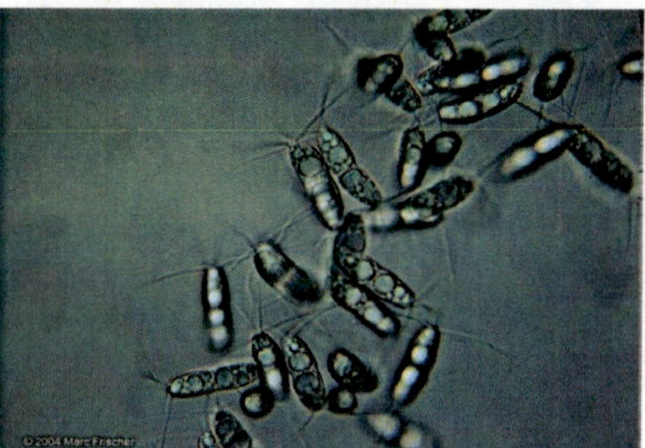

Fig. 72: Fruiting bodies and ascospores of *Corollospora matrima*

2.5.17. Halosphaeria species

Halosphaeria species (Fig.73) is *a* Marine fungi.

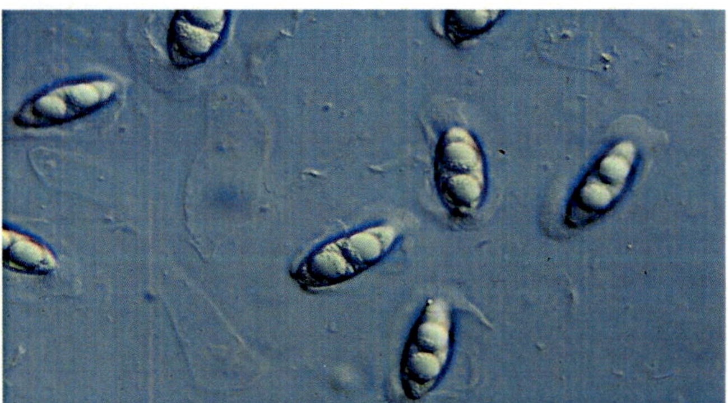

Fig. 73: Ascospores of *Halosphaerias* spp

2.5.18. Loramyces species

The genus *Loramyces*, an **ascomycete** from the class **Leotiomycetes**, contains two aquatic species, ***L. junicola*** (1), and ***L. macrosporus*** (2). Both species are **saprophytes** found **growing on submerged, decaying plants.**

Loramyces junicola: Its natural habitat is **freshwater ponds** (Fig.74).

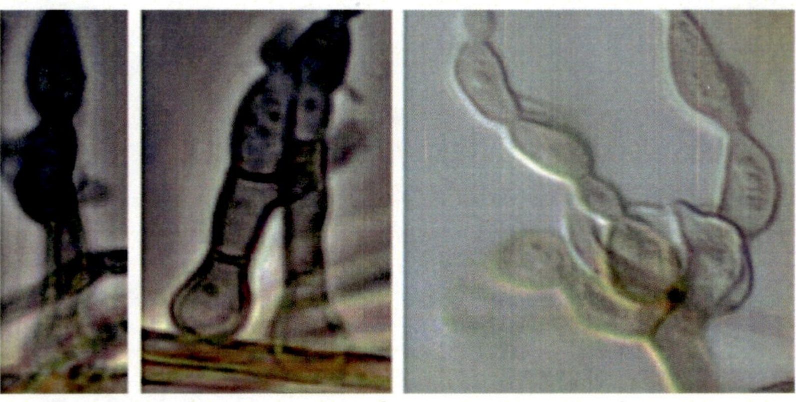

Fig. 74: Asci and Ascospores of *Loramyces junicola*

2.5.19. Nimbomollisia species

It is found in **Lentic** habitat (Fig.75).

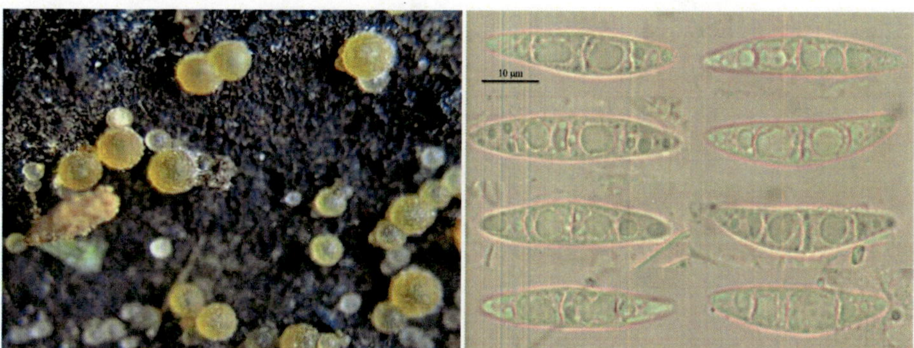

Fig. 75: Fruiting bodies and Ascospores of *Nimbomollisia* spp.

2.5.20. Obtectodiscus aquatius

It is a fresh water ascomycetes (Fig.76).

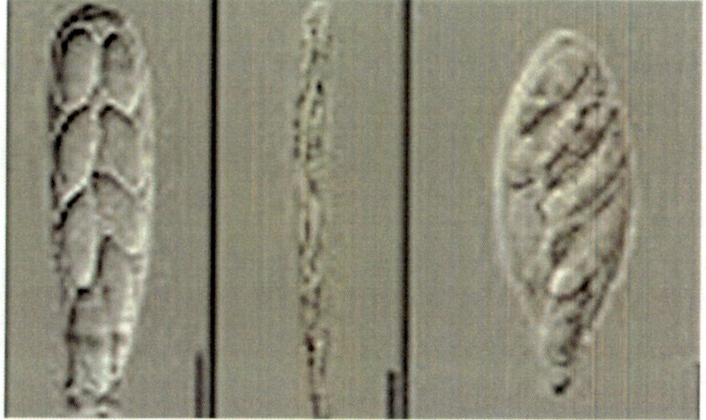

Fig. 76: Asci and Ascospores of *Obtectodiscus aquatius*

2.5.21. Fluviatispora **boothii**

It is a fresh water fungi (Fig.77).

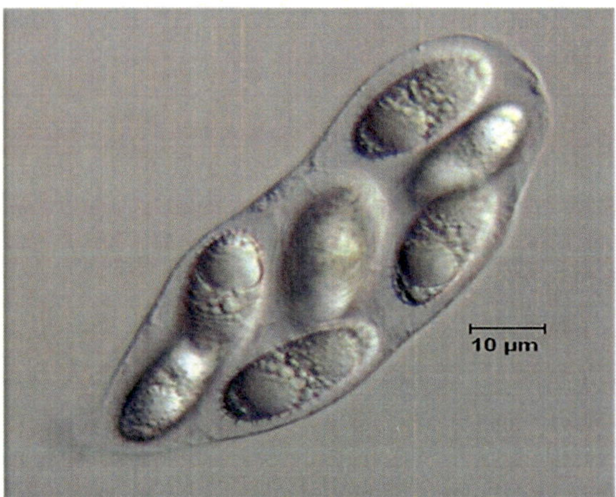

Fig. 77: Asci and Ascospores of *Fluviatispora boothii*

2.5.22. Caryospora species

It is a fresh water ascomycetes (Fig.78).

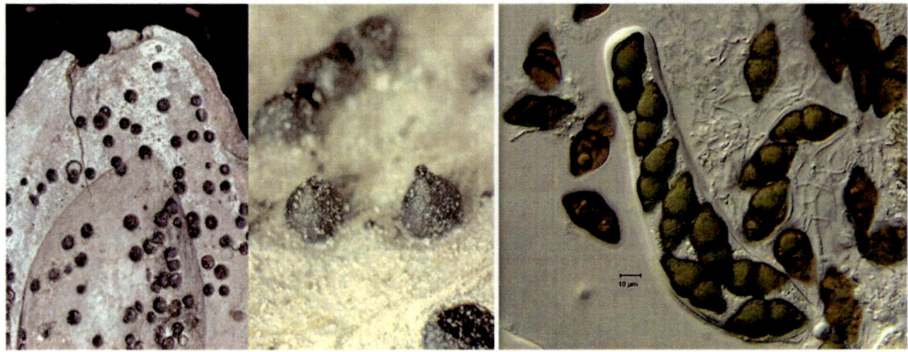

Fig. 78: Fruiting bodies and Ascospores of *Caryospora* spp.

2.5.23. Kirschsteiniothelia maritima

It is found **on wood, especially resin-coated coniferous wood, in the sea** (Fig.79).

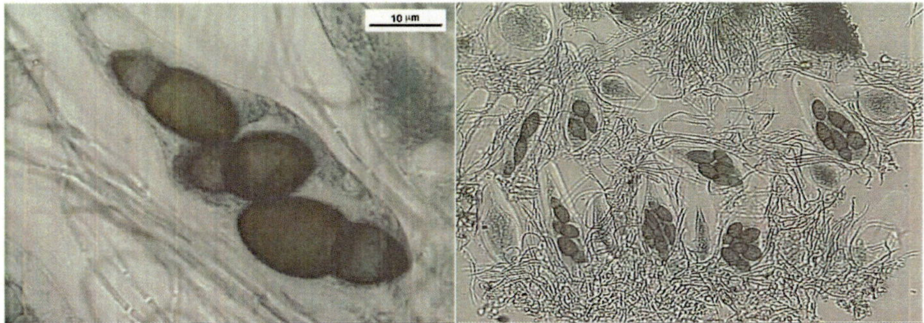

Fig. 79: Asci and Ascospores of *Kirschsteiniothelia maritima*.

2.5.24. Kirschsteiniothelia elaterascus

It is found in **fresh water habitat** (Fig.80).

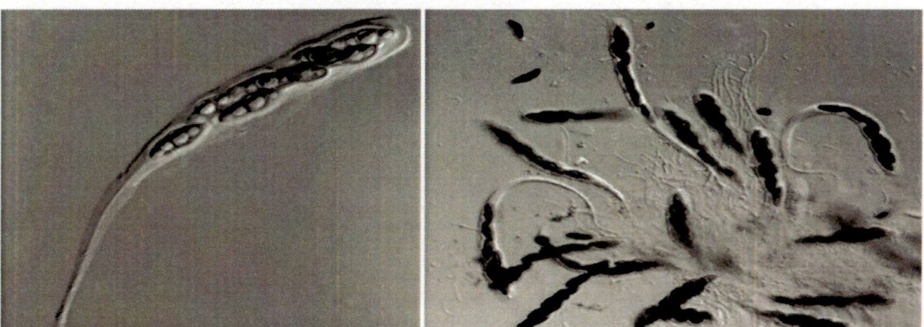

Fig. 80: Asci and Ascospores of *Kirschsteiniothelia elaterascus*.

2.5.25. Massarina species

Massarina (Fig.81) genus fungi are found in fresh water, terrestrial and marine habitat.

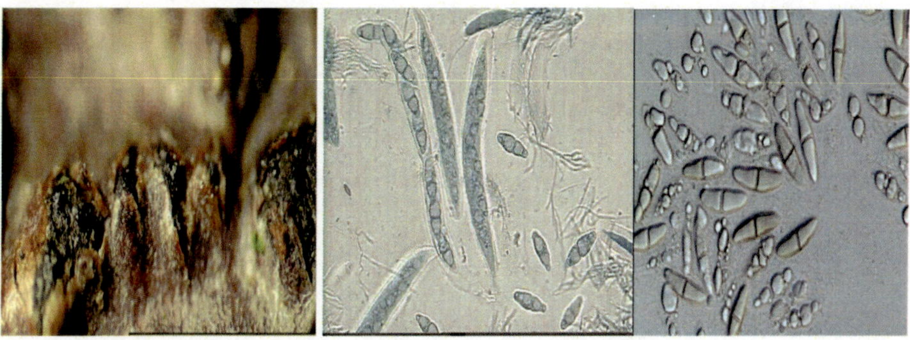

Fig. 81: Fruiting bodies, Asci and Ascospores of *Massarina* spp.

2.5.26. Phaeosphaeria species

The fungal genus is found in **Lentil** habitat (Fig.82).

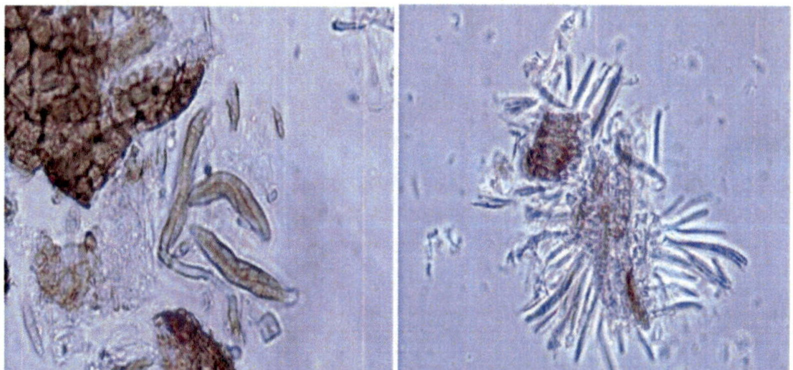

Fig. 82: Fruiting bodies, Asci and Ascospores of *Phaeosphaeria* spp

2.5.27. Pleospora species

It is found in **fresh** and **marine** habitat (Fig.83).

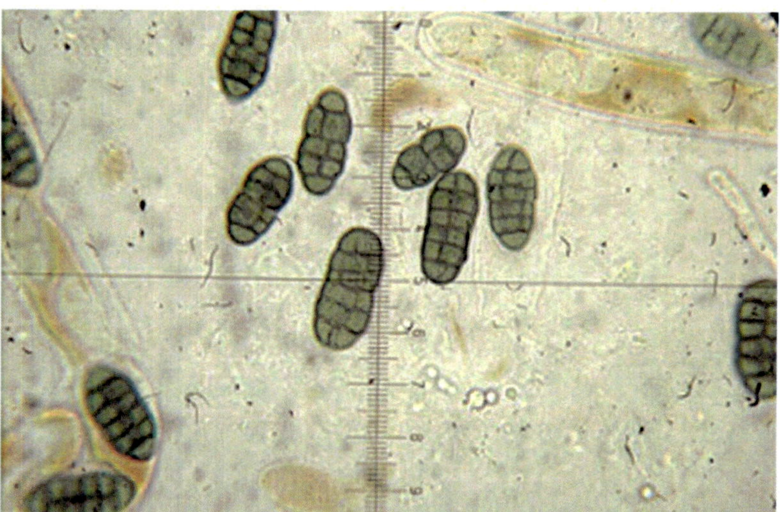

Fig. 83: Spores of *Pleospora* spp

2.5.28. Rebentischia species

It is an aquatic fungi (Fig.84).

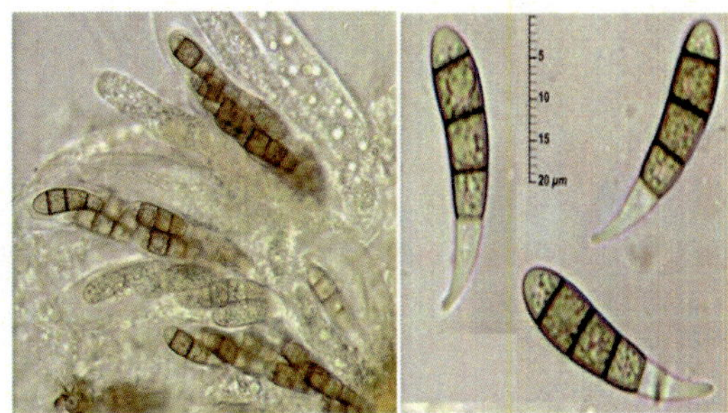

Fig. 84: Fruiting bodies and Spores of *Rebentischia* spp

2.5.29. Jahnula species

It is found in fresh water habitat (Fig.85).

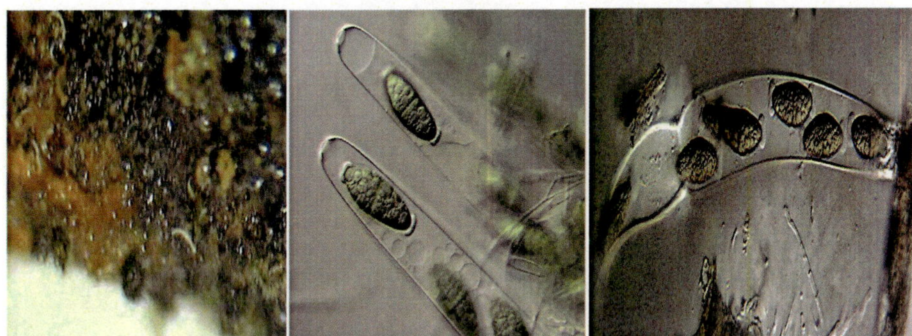

Fig. 85: Fruiting bodies, Asci and ascospores of *Jahnula* spp

2.5.30. Lophiostoma species-

It is found in **Fresh water colonizing dead culm tissues** (Fig.86).

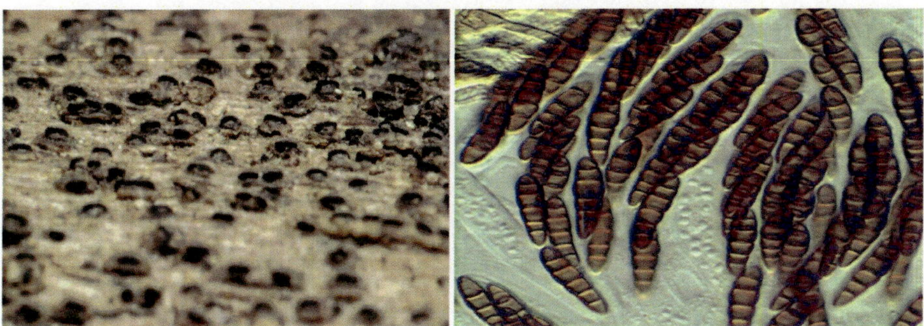

Fig. 86: Fruiting bodies and spores of *Lophiostoma* spp

2.5.31. Halosarpheia spp.

Halosarpheia spp (Fig.87) are found in fresh water, marine and brackish habitat.

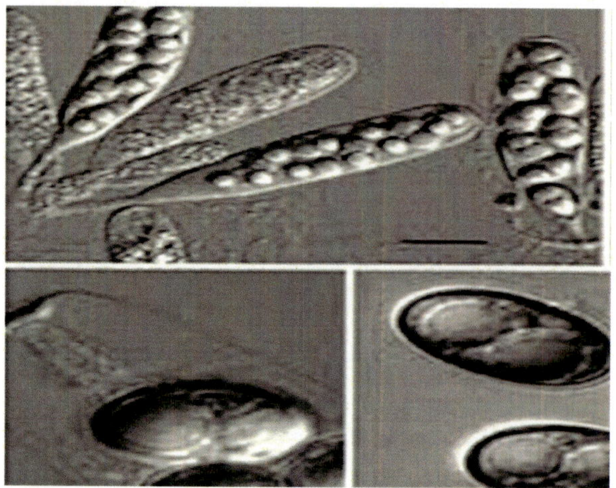

Fig. 87: Asci and Ascospores of *Halosarpheia* spp

2.5.32. Submersisphaeria species

It is found in **fresh water** habitat (Fig.88).

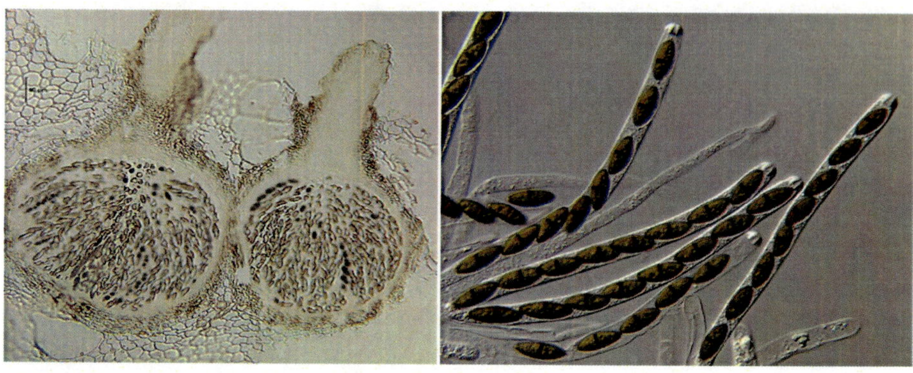

Fig. 88: Fruiting bodies, Asci and Ascospores of *Submersisphaeria* spp

2.5.33. Ophioceras species

Ophioceras species (Fig.89) are found on **decaying terrestrial palms, millipedes and woody substrates in freshwater habitats.**

Fig. 89: Spores of *Ophioceras* spp in infected Millipedes

2.5.34. Acrogenospora spp

It is found on submerged wood in fresh water ecosystem (Fig.90).

Fig. 90: Fruiting bodies of *Acrogenospora* spp.

2.5.35. Beltrania rhombic

It is found on **submerged wood in fresh water ecosystem** (Fig.91).

Fig. 91: Fungal spores of Beltrania rhombic

2.5.36. Sporoschisma spp.

It is found on **submerged wood** (Fig.92).

Fig. 92: Fungal spores of *Sporoschisma* spp

2.5.37. *Leptosphaeria species*

It is found **on the dead leaves of macrophytes in aquatic habitat** (Fig.93).

Fig. 93: Fruiting bodies and spores of *Leptosphaeria* spp

2.5.38. Nodulosphaeria species

It is found **on the dead leaves of macrophytes in aquatic habitat** (Fig.94).

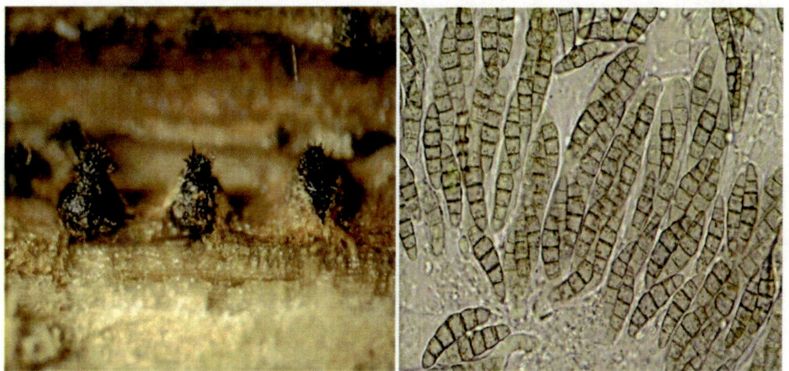

Fig. 94: Fruiting bodies and spores of *Nodulosphaeria* spp

2.5.39. Paraphaeosphaeria species

It is found **on the dead leaves of macrophytes in aquatic habitat** (Fig.95).

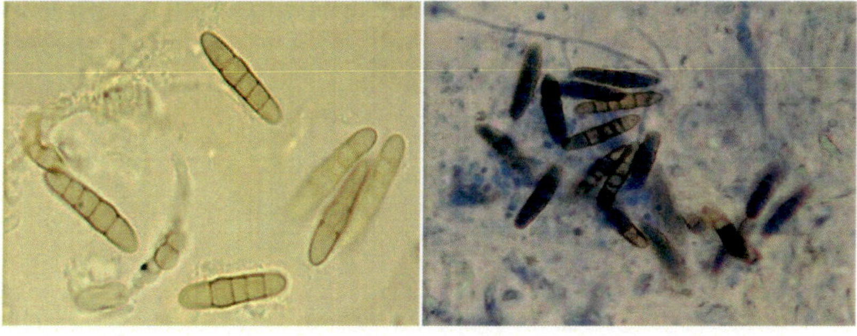

Fig. 95: Fungal spores of *Paraphaeosphaeria* spp

2.5.40. Phaeosphaeria species

Phaeosphaeria species (Fig.96) is found on the dead leaves of macrophytes in aquatic habitat.

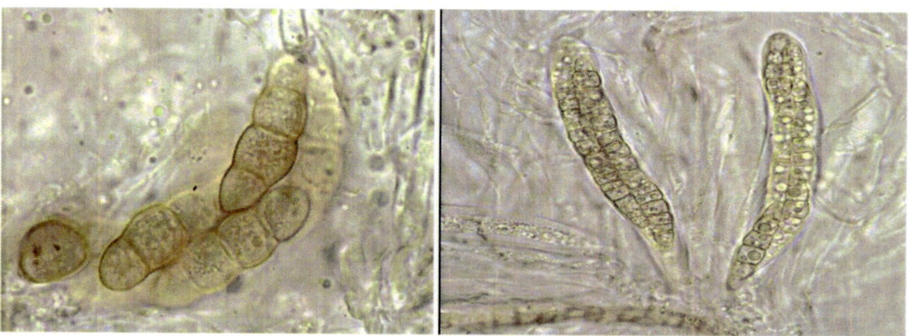

Fig.96: *Fungal spores of Phaeosphaeria species*

2.5.41. Nectria species

Nectria **species** (Fig.97) are found in **freshwater ecosystems. These are also regularly observed developing on old fruiting bodies of various ascomycetes on submerged wood, and these may be mycoparasites.**

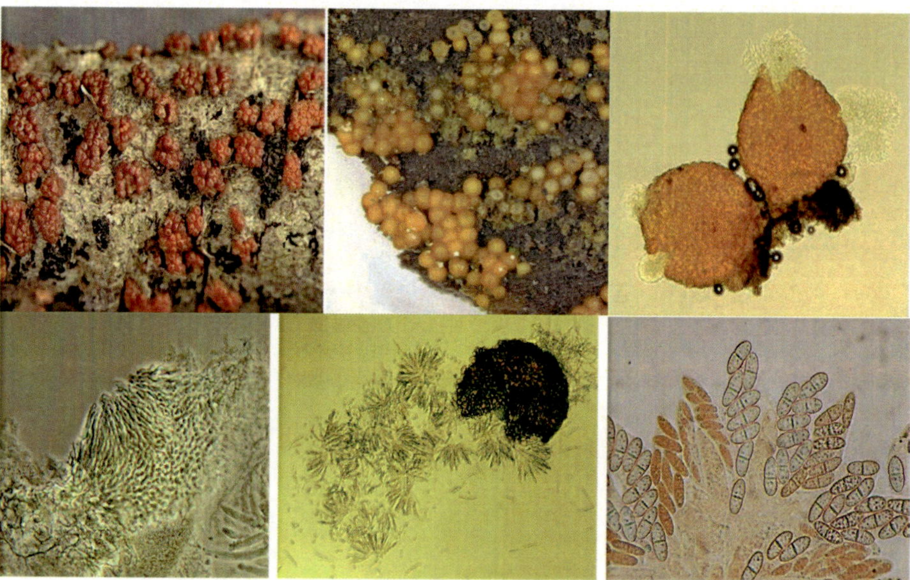

Fig.97: Fruiting bodies and fungal spores of *Nectria* species

2.5.42. Achlya species

Achlya (Fig.98) is a genus of **water mould,** regularly reported **growing in association with diseased fish and can also parasitise captive aquatic fauna such as turtles, tadpoles and fish.**

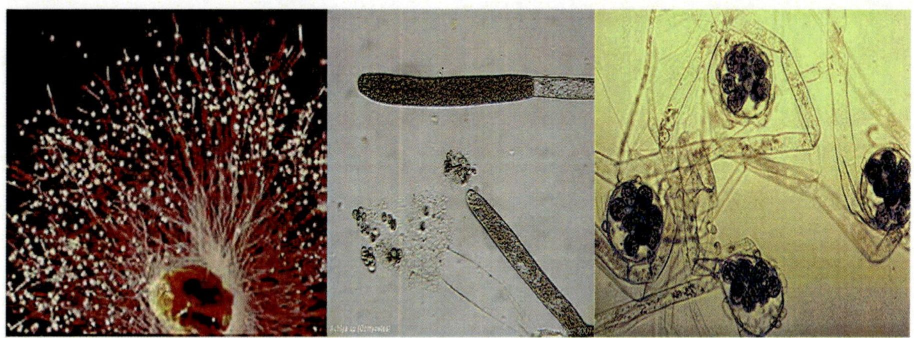

Fig. 98: Fruiting bodies and fungal spores of *Achlya species*

2.5.43. Aphanomyces astaci

Aphanomyces astaci (Fig. 99), is a water mold that infects crayfish.

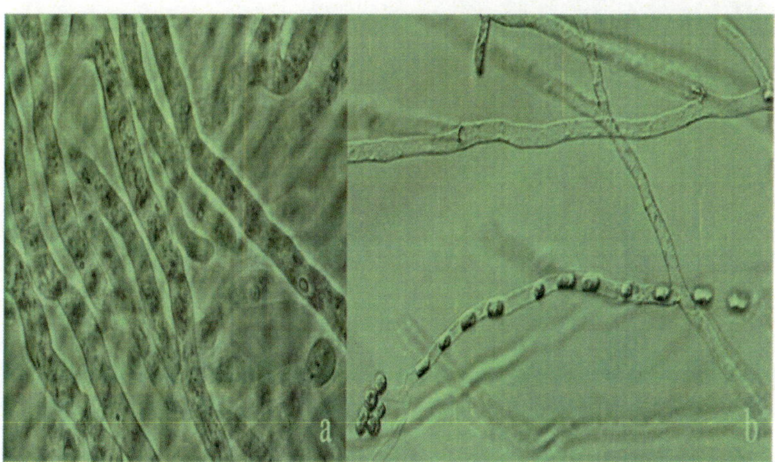

Fig. 99: Fungal spores of *Aphanomyces astaci*

2.5.44. Pythium species

Pythium (Fig.100) is a genus of **water mould,** regularly reported **growing in association with diseased fish and can also parasitise captive aquatic fauna such as turtles, tadpoles and fish.**

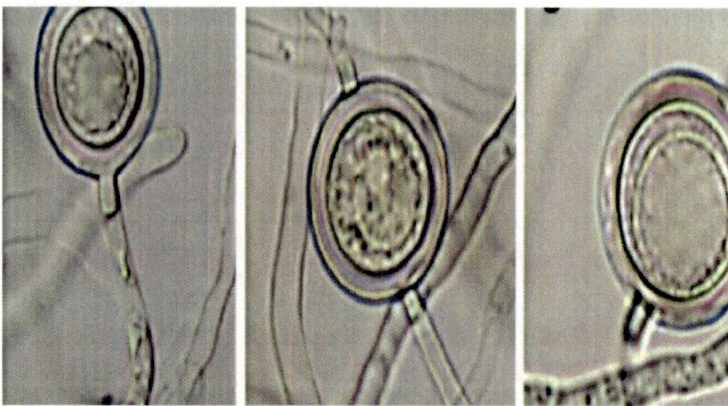

Fig.100: Fungal spores of *Pythium species*

2.5.45. Saprolegnia species

A common water mold (Fig.101) found in marine habitat and stagnated water.

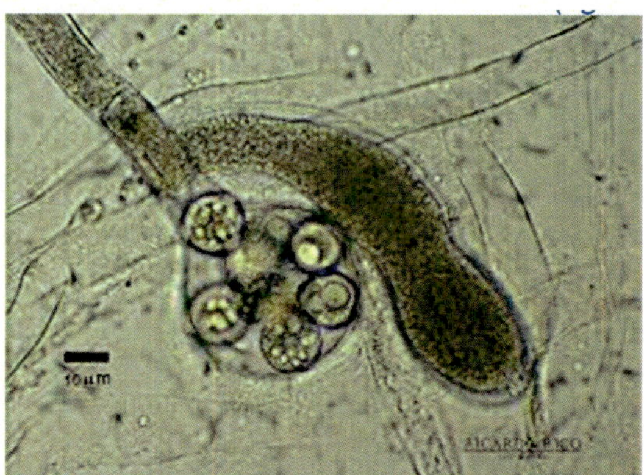

Fig.101: Fungal spores sac and spores of *Saprolegnia species*

2.5.46. Absidia species *(Fig.102)*

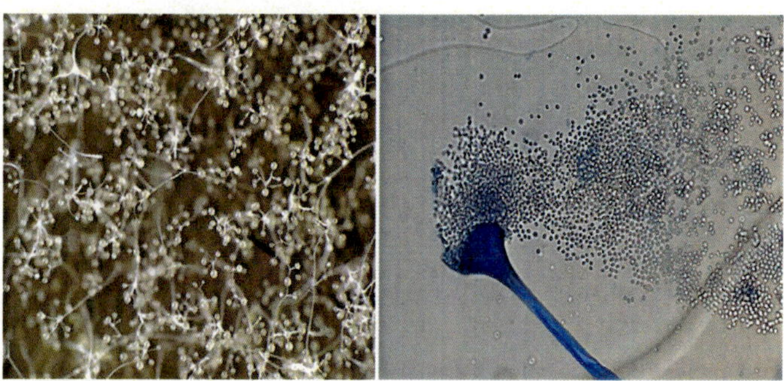

Fig.102: Fungal spores sac and spores o*f Absidia sp.*

2.5.47. Acremonium species (Fig.103)

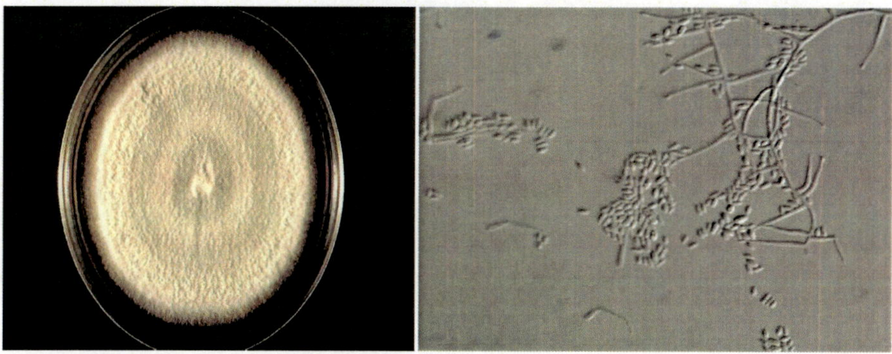

Fig.103: Fungal growth and spores of *Acremonium* sp

2.5.48. Bipolaris species (Fig.104)

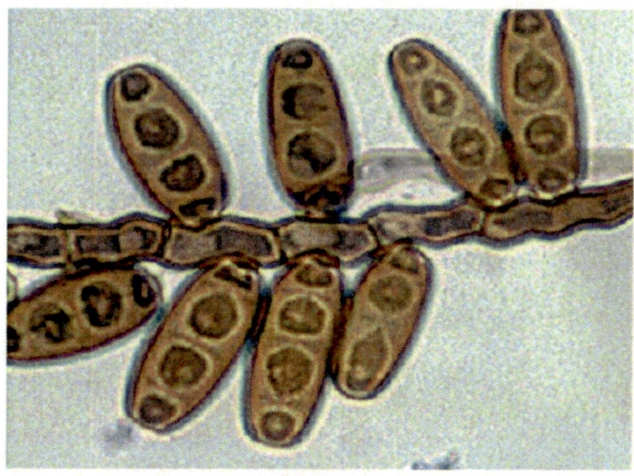

Fig.104: Fungal spores of *Bipolaris* sp

2.5.49. Blastomyces dermatidis

The fungus Blastomyces dermatidis (Fig.105) lives in **soil and wet, decaying wood, often in an area close to a waterway such as a lake, river or stream.** It is a **dimorphic fungus.**

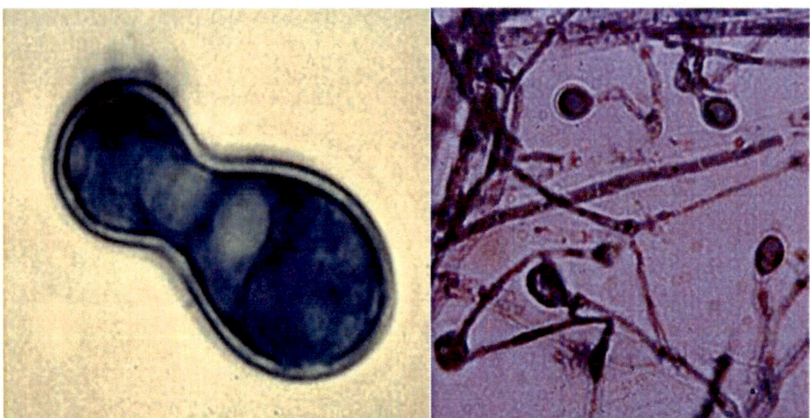

Fig.105: *Fungal spores of Blastomyces dermatidis*

2.5.50. Candida albicans (Fig.106)

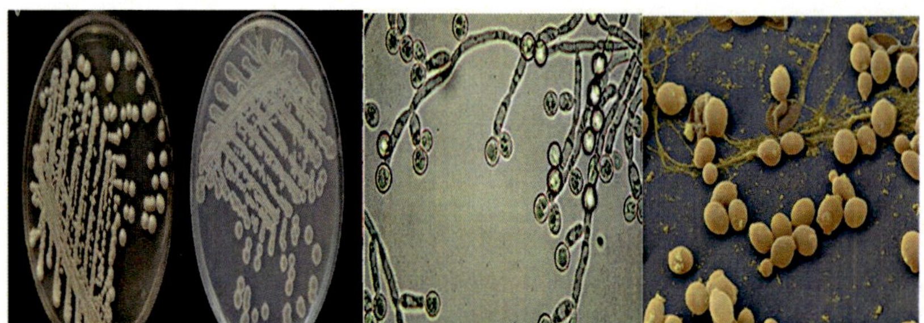

Fig.106: *Fungal growth and spores of Candida albicans*

2.5.51. Coccidioides species (Fig.107)

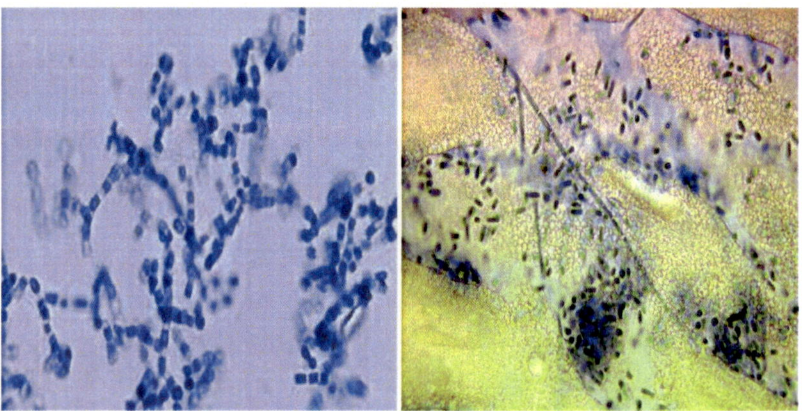

Fig.107: Fungal spores of *Coccidioides* sp.

2.5.52. Curvularia species (Fig.108)

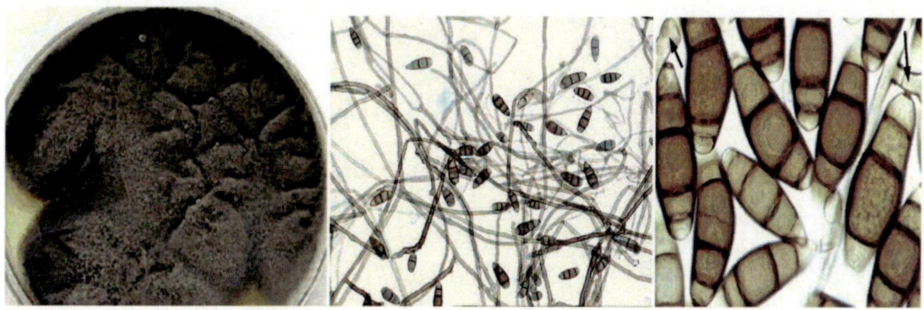

Fig. 108: Fungal growth and spores of *Curvularia* sp.

2.5.53. Epidermophyton floccosum (Fig.109)

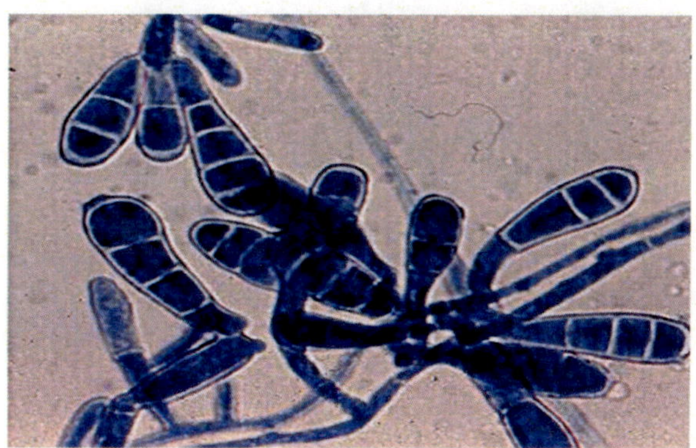

Fig. 109: Fungal spores of Epidermophyton floccosum.

2.5.54. Exophiala (Fig.110)

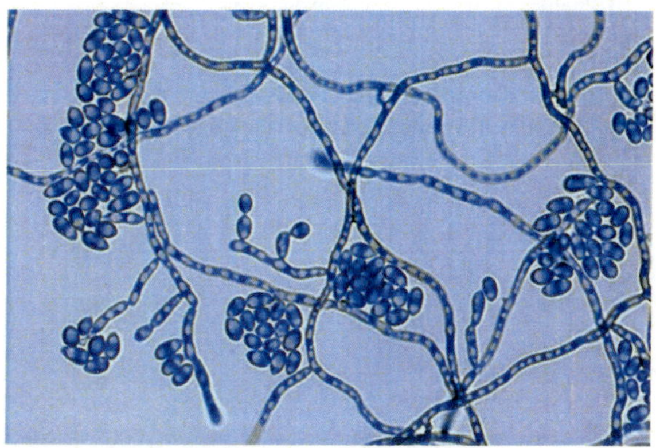

Fig. 110: Fungal spores of *Exophiala* sp.

2.5.55. Geotrichum (Fig.111)

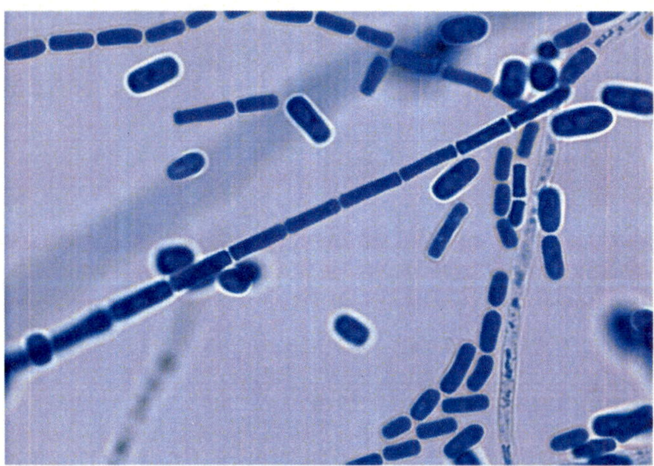

Fig. 111: Fungal spores of *Geotrichum* sp.

2.5.56. Histoplasma (Fig.112)

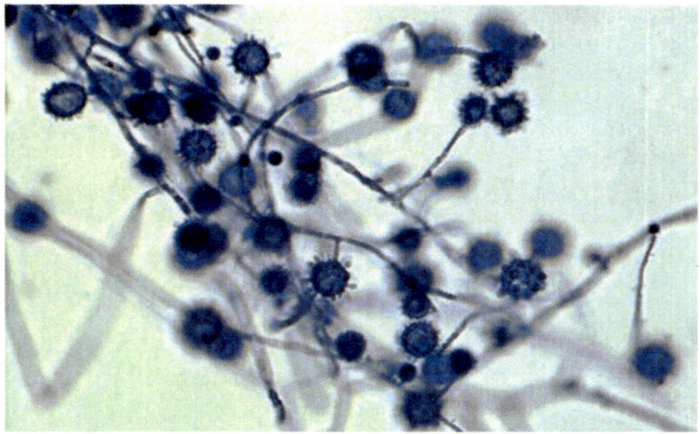

Fig. 112: Fungal spores of *Histoplasma* sp.

2.5.57. *Madurella* (Fig.113)

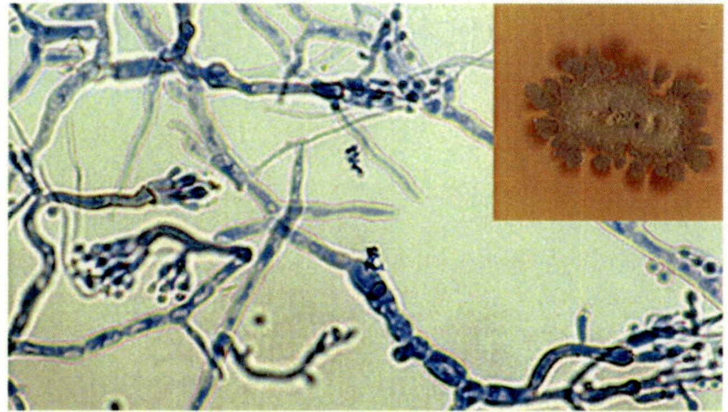

Fig. 113: Fungal growth and spores of *Madurella* sp.

2.5.58. Microsporum (Fig.114)

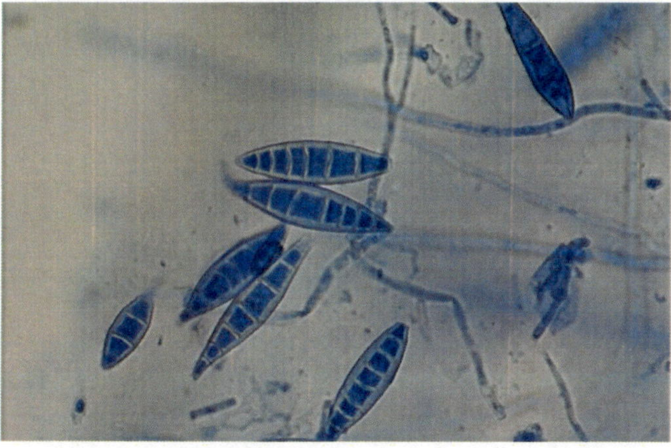

Fig.114: Fungal spores of *Microsporum* sp.

2.5.59. Neotestudina (Fig.115)

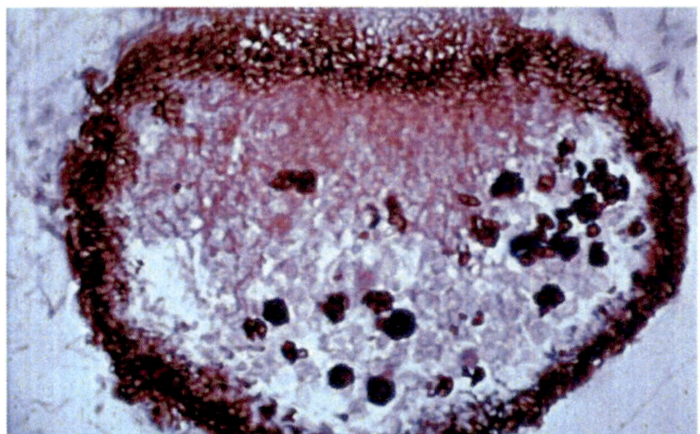

Fig. 115: Fungal fruiting body of *Neotestudina* sp.

2.5.60. Paracoccidioides brasiliensis (Fig.116)

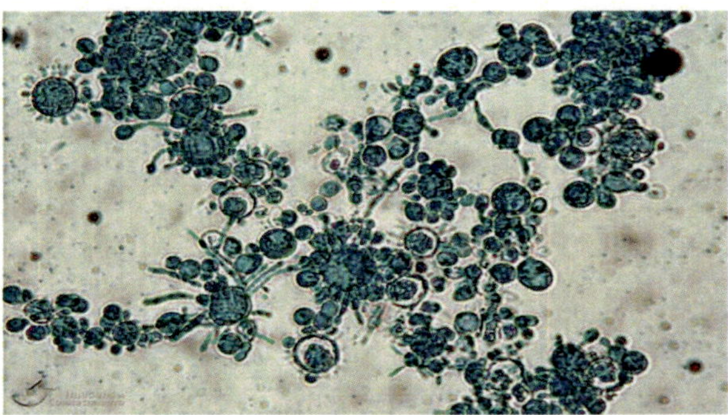

Fig. 116: Fungal spores of Paracoccidioides brasiliensis.

2.5.61. Piedraia hortae (Fig.117)

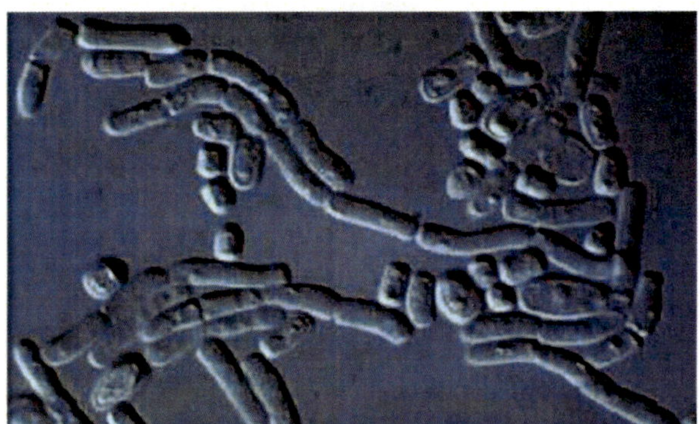

Fig. 117: Fungal spores of Piedraia hortae

2.5.62. Pseudallescheria boydii (Fig.118)

Fig.118. Fungal fruiting body and spores of *Pseudallescheria boydii*

2.5.63. Pyrenochaeta (Fig.119)

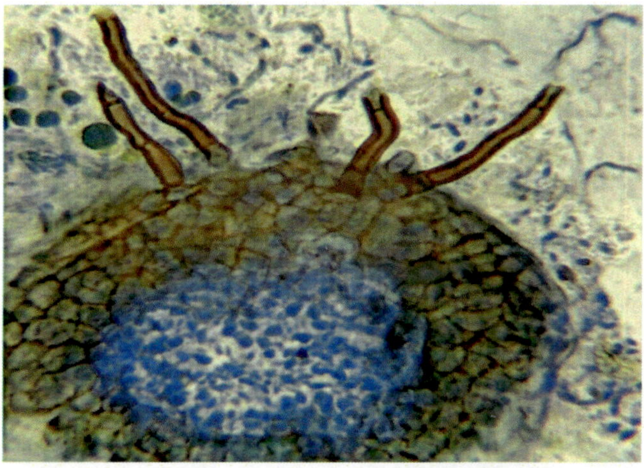

Fig. 119. Fungal fruiting body and spores of *Pyrenochaeta* sp.

2.5.64. *Rhodotorula (Fig.120)*

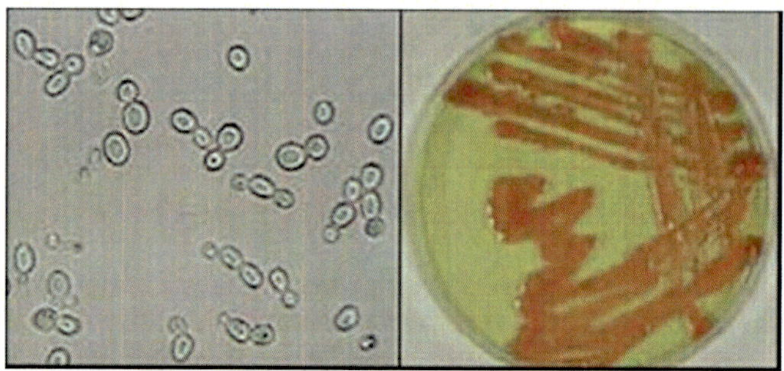

Fig. 120. Fungal growth and budding spores of *Rhodotorula* sp.

2.5.65. *Saccharomyces* (Fig.121)

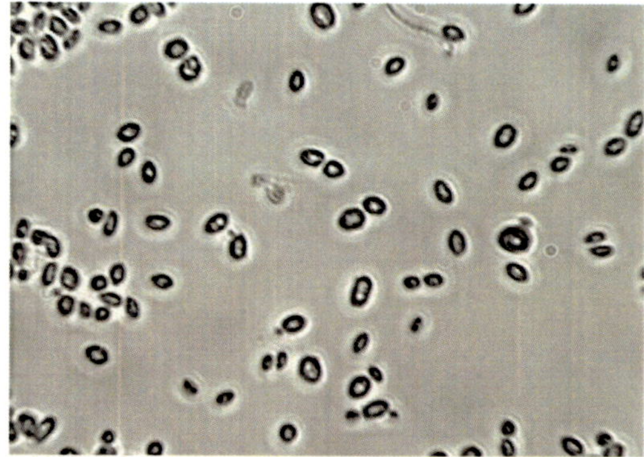

Fig. 121. Fungal spores of *Saccharomyces* sp.

2.5.66. *Sporothrix* (Fig.122)

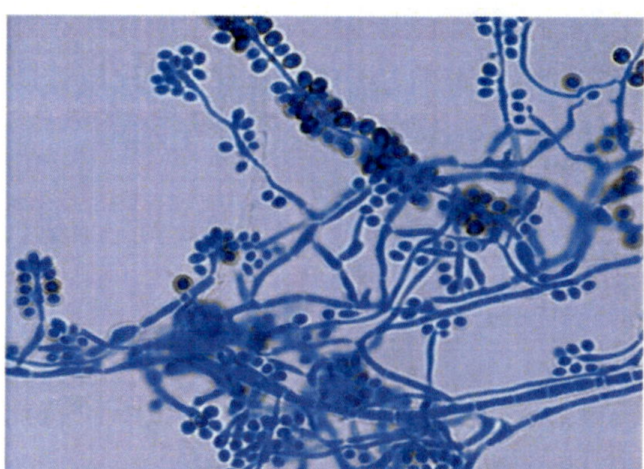

Fig. 122: Fungal structure and spores of Sporothrix sp.

2.5.67. *Trichosporon (Fig.123)*

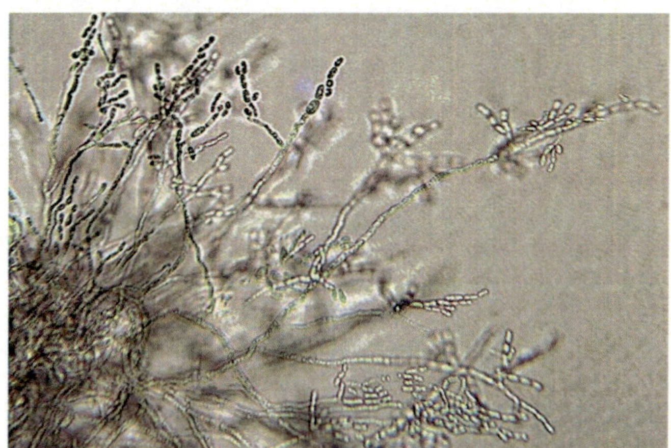

Fig. 123: Fungal structure and spores of *Trichosporon* sp.

2.5.68. ***Rhizopus*** **(Fig.124)**

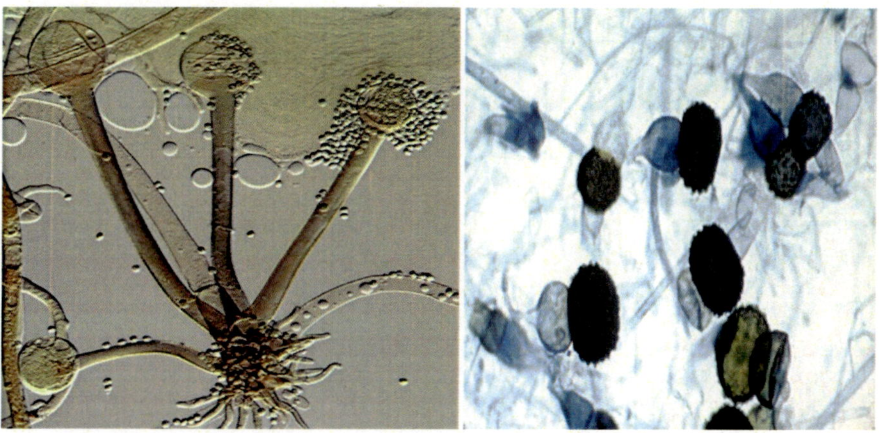

Fig. 124. Fungal structure and spores of *Rhizopus* sp.

Mitosporic fungi *like Aquaphila, Canalisporum, Camposporidium, Candelosynemma, Clohesyomyces, Conioscyphopsis, Elagantimyces, Enridescalsia, Jeranium, Nidulispora, Paracryptophiale, and Yinmingella* etc. **can be classified as indwellers because they have been reported only from freshwater habitats.**

Acrodictys, Acrogenospora, Arthrobotrys, Bactrodesmium, Berkleasmium, Brachysporiella, Brachysporium, Camposporium, Cordana, Conioscypha,

Dactylaria, Dictyosporium, Ellisembia, Exserticlava, Monodictys, Neta, Spadicoides, Sporidesmiella, Sporoschisma, Taeniolella, Trichocladium, Vanakripa, and Xylomyces **can be classified as immigrants because they are reported from terrestrial as well as freshwater habitats.**

Species common in Australian upland stream included *Tetrachaetum elegans, Lunulospora cymbiformis, Flagellospora penicillioides* and *Alatospora acuminate.*

Fungal genera commonly found in wetlands include *Alternaria, Cylindrocarpon, Cladosporium,Penicillium, Fusarium, Trichoderma* and aquatic hyphomycetes (*Alatospora, Tetracladium,Helicodendron, Helicoon, Crucella subtilis, Janetia curviapicis* was described as ``associated with or growing on other hyphomycetes on submerged wood" from a stream in north Queensland, Australia (Goh and Hyde, 1996), while the *Crucella subtilis* anamorph of *Camptobasidium hydrophilum* is apparently a mycoparasite of several aquatic hypho-mycetes (Marvanova and Suberkropp, 1990). Semi-submerged leaves of water lilies may also be infected by the chytridiomycete *Physoderma limnanthemi*

Section II
Fungi Harmful to Living Being

3

Fungi Causing Disease in Plant and Food Product

3.1. Plant diseases caused by soil borne fungi

3.1.1. Clubroot Disease causing fungi

Plasmodiophora brassicae (Fig.125)

It infects Crucifers host species.

Infected plants in the begening shows pale green to yellowish leaves, which later on show wilting in the middle of hot, sunny days, and recovering during the night. Young plants may be killed by the disease soon after infection, whereas older plants may remain alive but become stunted and fail to produce marketable heads.

The most characteristic symptoms of the disease appear on the roots as spindle-like, spherical, knobby, or club-shaped swellings. The swellings may be few and isolated or they may coalesce and cover the entire root system. The older and usually the larger clubbed roots disintegrate before the end of the season because of invasion by bacteria and other fungi.

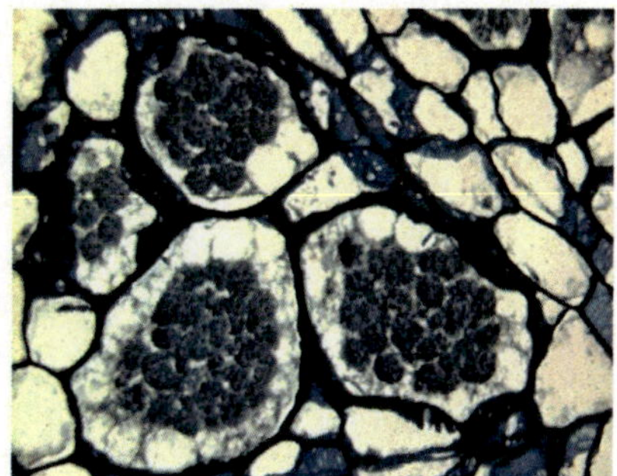

Fig. 125: Clubroot disease symptoms and its pathogen

3.1.2. Root Rot Disease causing fungi

***Pythium* spp, *Phythophthora* spp, *Monosporascus cannonballus, Gaeumannomyces (Ophiobolus) graminis, Armillaria* spp, *Phellinus* spp, *Phymatotrichum* spp, *Sclerotium* spp etc.**

These fungal pathogens (Fig.126) infect pepper, cucurbits, tomato, strawberry, citrus, cacao, apple, peach, cherries, melons, cotton and forest trees.

The infected plants above ground parts show symptoms of wilting, stunting, yellowing, necrosis, sparse foliage and branch dieback. In many cases the symptoms get progressively worse until the plant dies. A common symptom is a gradual fading in the colour of the foliage, from a vibrant to a dull green, through to greyish and finally brown.

The examination of the roots, collar and stem base of an affected plant will reveal a poor root system. Many of the fine, feeder roots get rotted away. Some or all of the larger roots also show evidence of decay and turn brown or black internally, softer than normal and may break easily. The fruits on the infected plants fails to progress toward ripening properly and becomes unmarketable.

The roots of affected plants show lesions, especially at root junctions, the feeder and secondary roots decay and slough off, and under wet conditions, more roots rot and the tap root is killed.

Fig. 126: Rootrot disease symptoms and its pathogens

3.1.3. Damping Off Disease causing fungi

Pythium, Phytophthora, Rhizoctonia, Thielaviopsis, Fusarium, Sclerotium spp, *and Sclerotinia* spp.

These fungal pathogens (Fig.127) infect tobacco, chillies, tomato, cucurbits, beans, tomato, coriander, spring onions, beetroots, eggplant, brassicas, and leafy vegetables.

The first evidence of damping-off or seed piece decay (as in potatoes) is the failure of some plant seedlings to emerge. If seeds are attacked before they germinate, they become soft and mushy, turn dark brown, and decay. They may have a layer of soil clinging to them when they are dug up because the soil is interwoven with fine, threadlike fungus growth. Germinating seedlings shrivel and may darken. If seedlings are attacked after they emerge, stem tissue near the soil line start decay and weakened, usually causing plants to topple and die. When only roots are decayed, plants may continue standing but remain stunted, wilt and eventually die. As seedlings get older, they become less susceptible to damping-off pathogens.

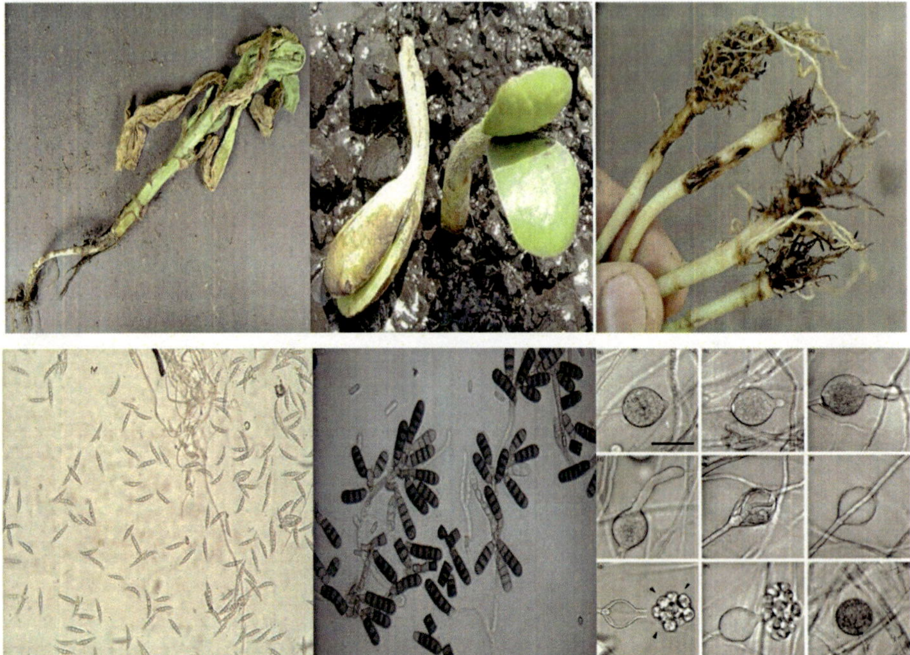

Fig. 127: Damping off disease symptoms and its pathogens

3.1.4. Wilt Disease causing fungi

Fusarium, Verticilium, Ceratocystis, Ophistoma

These fungal pathogens (Fig.128) infect host plants like tomato, banana, vegetables, flowers, fruit trees, strawberries, field crops, and shade and forest trees.

The first disease symptom appears as slight vein clearing on the outer, younger leaflets. Subsequently, the older leaves show epinasty caused by drooping of the petioles. Plants infected at the seedling stage usually wilt and die soon after appearance of the first symptoms. Older plants in the field may wilt and die suddenly if the infection is severe and the weather is favourable for the pathogen.

More commonly, in older plants, vein clearing and leaf epinasty are followed by stunting of the plants, yellowing of the lower leaves, occasional formation of adventitious roots, wilting of leaves and young stems, defoliation, marginal necrosis of the remaining leaves, and finally death of the plant. Often these symptoms appear on only one side of the stem and progress upward until the foliage is killed and the stem dies. Fruit may occasionally become infected and then it rots and drops off without becoming spotted. Roots get infected; during an initial period of stunting, the smaller roots rot.

In cross sections near the base of the infected plant stem, a brown ring is evident in the area of the vascular bundles. The upward extent of the discoloration depends on the severity of the disease. At higher magnification, tyloses can be seen inside vessels of newly infected shoots that block the upward movement of nutrients and water. The symptoms of *Verticillium* wilts are almost identical to those of Fusarium wilts. In many hosts and most areas, however, *Verticillium* induces wilt at lower temperatures than *Fusarium.*

Fig. 128: Wilt disease symptoms and its pathogen

3.1.5. Wart disease causing fungi

Synchytrium endobioticum.

It infects Potato plant and *causes black wart disease in potato (Fig.129).*

The wart symptoms can be found on all underground plant parts except roots. On stem, galls form at the base of the stem which appears initially white, but turning black when decaying. These galls may be as small as a pin or as large as a fist. The surface of galls is rough and corrugated –warty in appearance. On potato tubers, the eyes develop cauliflower-like swellings. When formed underground, they are of the same colour as the potato skin, darkening with age, or green if exposed to light. Typical warts are soft and pulpy and easier to cut than a healthy tuber. On stolen symptoms appears similar to tuber, while on aerial bud small greenish warts form in the position of the aerial buds at the stem bases.

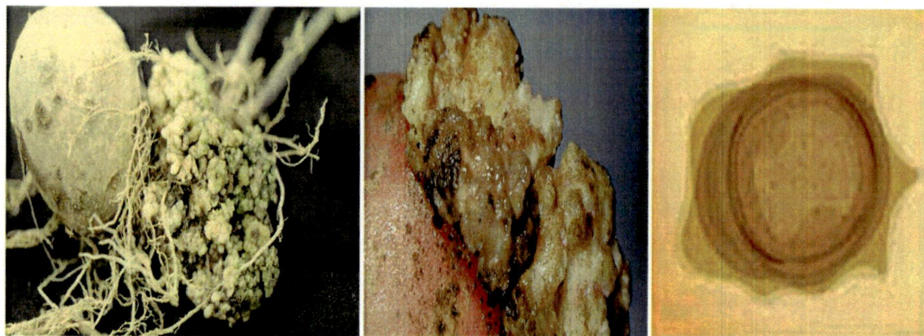

Fig. 129: Wart disease symptoms and its pathogen

Crown Wart of Alfalfa

It is caused by the fungus ***Physoderma alfalfa***

The fungus infects the Alfalfa plant (Fig.130)

The fungal pathogen causes large, knobby swellings on crown and roots. The plant becomes stunted, unthrifty, and often dies.

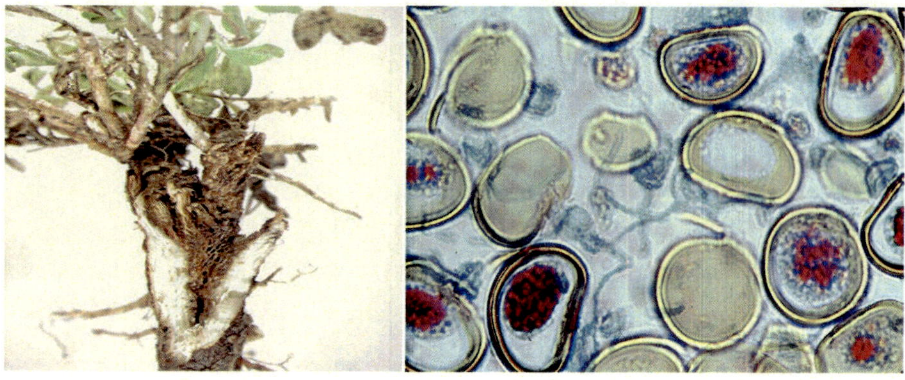

Fig. 130: Crown wart disease symptoms and its pathogen

3.1.6. Foot Rot disease causing fungi

Pythium aphanidermatum

The fungal pathogen infects Cereal crop (Fig.131), Papaya and other fruits plants host.

The symptoms of foot rot is characterised by the appearance of water soaked patches at the base of the stem near the soil. The disease appears during the rainy season and becomes severe when the rain fall is high and temperature is in the range of 30-36ºC. The wet patches on the basal stem enlarge rapidly and girdle the stem. The infected portions turn brown to black due to rotting of the parenchymatus tissues. Such plants are unable to with stands gusts of winds and topple down and die. If only one side of the stem is rotted, the plants survive, but not for long. Their growth is stunted and fruits, if formed, are shrivelled and malformed.

Fig.131: Foot rot disease symptoms and its pathogen

3.1.7. Koleroga disease causing fungi

Phytophthora palmivora

The fungal pathogen infects Arecanut Palm Tree (Fig.132)

Nuts are the main target of attack. Younger nuts are more prone to the infection. Nuts older than six months are resistant. During initial phase of infection, water soaked areas appear at the base of the nuts. The green colour of the rind becomes dark green. The infection gradually spreads over the entire nut, which finally gets detached and falls to the ground. In severe infections, the nuts are covered with white mycelium. Sometimes the entire crown of the tree is affected, resulting in withering of the leaves and young branches.

Fig. 132: Koleroga disease symptoms and its pathogen

3.1.8. Cigar End Rot disease causing fungi

Verticillium theobromae

The fungal pathogen infects banana fingers (Fig.133).

The symptoms include the necrosis of the fruit which starts from the tip (pistillate end) and progresses backwards. The banana fruit skin shrivel with blackened end of the fruit having powdery greenish conidia. The shrinking and withering skin looks like the burnt end of a cigar. Hence the name 'cigar end rot'. The internal rot in the fruit in the field is a 'dry rot'.

Fig. 133. Cigar end rot disease symptoms and its pathogen

3.1.9. Red Rot disease causing fungi

***Glomerella tucamanensis* (anamorph *Colletotrichum falcatum*)**

The fungal pathogen infects Sugarcane Crop (Fig.134).

Practically all above ground parts are affected, but stalks (stems) are the main target of attack. Symptoms may not be apparent in the early stages.

Plant growth is retarded and the canes get shrivelled; the rind shrinks and longitudinal wrinkles are formed.

The diagnostic symptom is the formation of red streaks, interrupted by white patches, inside the stalk. This can be seen by splitting the stalks longitudinally. The red patches are characteristic of the disease, not found in other stalk rots. The affected stalks show considerable loss in weight. The juice of the infected stalks emits unpleasant smell due to conversion of glucose to alcohol. Red longitudinal lesions are also seen on the mid rib of the leaves on which black acervuli of the anamorph, *Colletotrichum falcatum*, are formed.

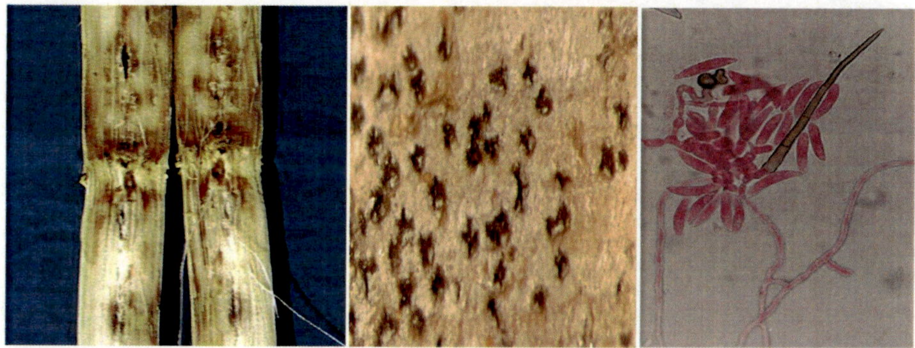

Fig. 134: Red rot disease symptoms and its pathogen

3.1.10. Dry Rot disease causing fungi

Rhizoctonia **spp.**

The fungal pathogen infects cotton, sorghum, pulses and oilseed plants (Fig.135) etc.to cause this disease.

The disease appears in the form of wilting and drying of plant parts. The affected roots show browning of the root portions. The dried plants can not be uprooted easily. The affectd roots remain dried.

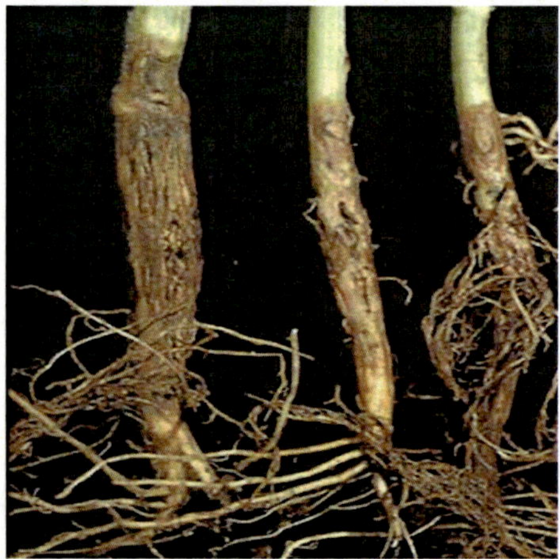
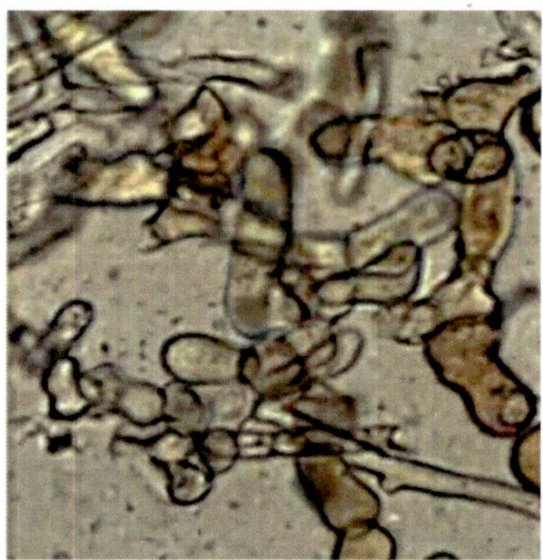

Fig. 135: Dry root rot disease symptoms and its pathogen

3.2 Plant diseases caused by air borne fungi

3.2.1. Downy Mildew disease causing fungi

Fungal species of genus ***Plasmopara, Bremia, Peronospora, Pseudoperonospora, Bremiella, Sclerospora etc.***

The downy mildew fungus infects Grapes, Crucifers, Vegetables, Soybeans, Sunflowers, Cucurbits, Tobacco and several others crops as its host.

The disease symptoms (Fig.136) in the beginning appears as small, pale yellow, irregular spots on the upper surface of the leaves, and a white downy growth of the sporangiophores of the downy pathogen appears on the underside of the spots. Later, the infected leaf areas are killed and turn brown, while the sporangiophores of the pathogen turn gray. The spots often enlarge, coalesce to form large dead areas on the leaf, and frequently result in premature defoliation.

Infected fruits particularly grapes are quickly covered with the downy growth, may become distorted or thickened, and may die. If infection takes place after the berries are half-grown, the fungal pathogen grows mostly internally; the berries become leathery and somewhat wrinkled and develop a reddish marbling to brown coloration. In case of late or localized infections of shoots, the shoots usually are not killed but show various degrees of distortion.

Fig. 136: Downy mildew disease symptoms on leaf and grape berries

Different genus of downy mildew fungi (Fig.137) are responsible for causing downy mildew in different crop species.

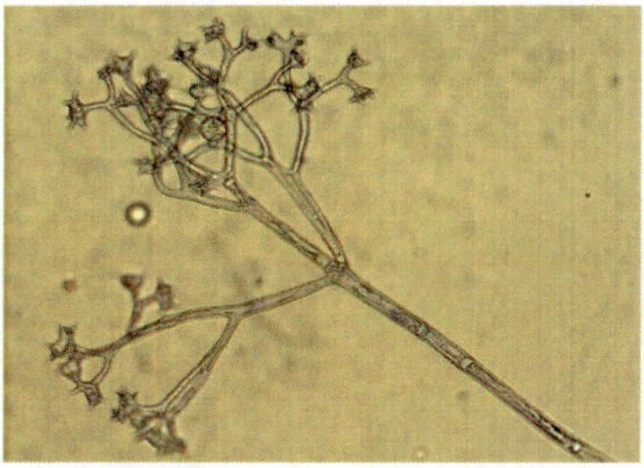

I. ***Plasmopara***

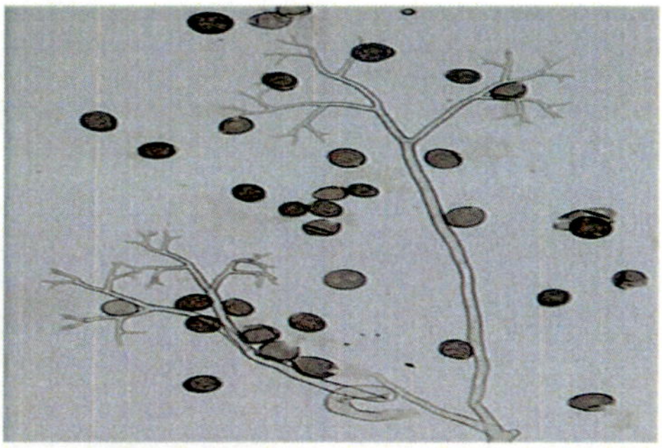

II. ***Sclerospora***

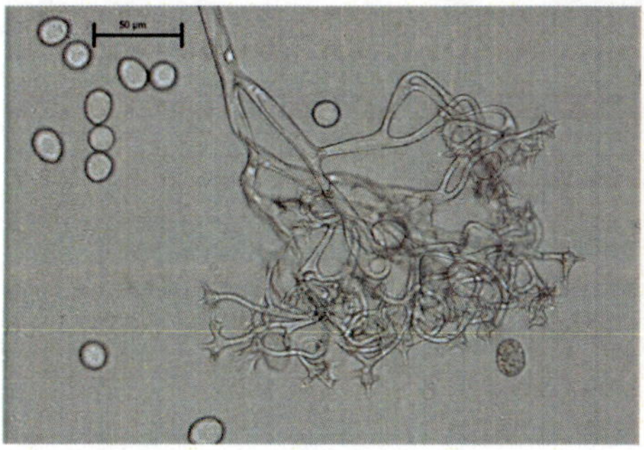

III. ***Bremia***

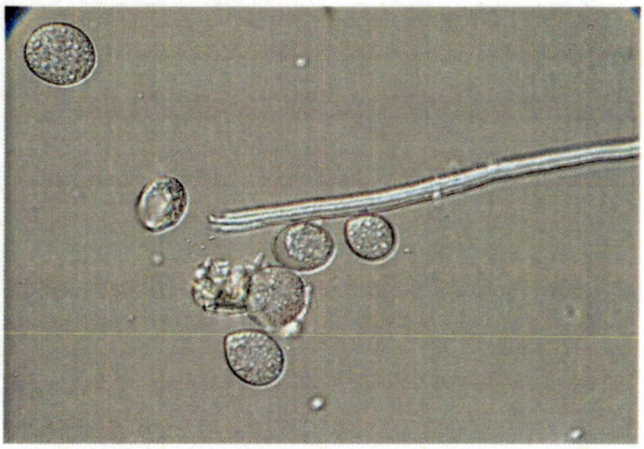

IV. ***Peronospora***

Fig. 137: Different genus of downy mildew fungi responsible for this disease

3.2.2. Powdery Mildew disease causing fungi

Species of ***Erysiphe, Podosphaera, Oidium, Sphaerotheca, Leveillula, Blumeria, Microsphaera, Phyllactinia etc.***

The fungus infects Cereals, Grapes, Legumes, onion, apple, pears, gourds and melons, Cucurbits, Crucifers etc.to cause powdery mildew disease (Fig.138).

Powdery mildews are probably the most common, conspicuous, widespread, and easily recognizable plant diseases. They affect all kinds of plants except gymnosperms.

The disease powdery mildews appear as spots or patches of a white to greyish powdery growth on young plant leaf tissues or on entire leaves and other plant parts sometimes being completely covered by the white powdery mildew growth. Powdery mildew is most common on the upper side of leaves, but it also affects the underside of leaves, young shoots and stems, buds, flowers, and young fruit.Tiny, pinhead-sized, spherical, at first white, later yellow-brown, and finally black fruiting bodies of fungi called cleistothecia may be present singly or in groups on the white to greyish mildew growth in the older areas of infection.

Fig.138: Powdery mildew disease symptoms and its pathogen

3.2.3. Rust disease causing fungi

Species of ***Puccinia, Uromyces, Hemileia, Cronartium, Melampsora, Phragmidium, Albigo, and Gymnosporangium etc***

The fungal pathogen infects Cereals, Legumes, apple, coffee, rose, flax, beans, crucifers, maize, pine, onions, celery etc.

The infection of the fungal pathogen leads to the formation of raised Pale leaf spots which eventually develop into spore-producing structures called pustules. The pustules are found most commonly on the lower leaf surface and produce huge numbers of microscopic spores (Fig.139). Pustules can be orange, yellow, brown, black or white. Some are a rusty brown colour, giving the disease its common name. In some cases, there may be dozens of pustules on a single leaf. Severely affected leaves often turn yellow and fall prematurely. Pustules also sometimes form on leaf stalks (petioles), stems and, rarely, on flowers and fruit. Heavy infection often reduces the vigour of the plant. In extreme cases (e.g. with antirrhinum rust) the plant can be killed.

Fig. 139: Rust disease symptoms and its pathogens

3.2.4. Smuts disease causing fungi

Ustilago

The fungal pathogen infects Cereals and Grasses (Fig.140).

When young seedlings are infected, minute galls form on the leaves and stems, and the seedling may remain stunted or may be killed. On older plants, infections occur on the young, actively growing tissues of axillary buds, individual flowers of the ear and tassel, leaves, and stalks. Infected areas are permeated by the fungus mycelium, which stimulates the host cells to divide and enlarge, thus forming galls. Galls are first covered with a greenish white membrane. Later, as the galls mature, they reach a size from 1 to 15 centimeters in diameter, and their interior darkens and turns into a mass of powdery, dark olive-brown spores. The silvery gray membrane then ruptures and exposes the millions of sooty teliospores, which are released into the air. Galls on leaves frequently remain very small (about 1–2 cm in diameter), hard, dry, and do not rupture.

Fig. 140: Smut disease symptoms and its pathogen

Loose Smut Disease

This disease generally appears on Oats, Barley, and Wheat plants (Fig.141).

Loose smut generally does not produce discernible symptoms until the plant has produced a head. Smutted plants sometimes head earlier than healthy ones, and smutted heads are often elevated above those of healthy plants.

In an infected plant, usually all the heads and all the spikelets and kernels of each head are smutted, i.e., they are transformed into a smut mass consisting of olive-green spores. Smutted kernels are at first covered by a delicate greyish membrane, which soon bursts and sets the powdery spores free. The spores are then blown off by the wind and leave the rachis as naked stalk.

Fig. 141: Loose Smut disease symptoms and its pathogen

3.2.5. Bunt disease causing fungi

Tilletia **sp.**

Tilletia caries (= T. tritici) and ***T. laevis (= T. foetida)*** cause the common bunt, whereas ***T. controversa*** causes dwarf bunt and ***T. indica*** causes Karnal bunt

The pathogen mostly infects the Wheat crop (Fig.142).

Plants infected with the common bunt fungi are usually a few to several centimeters shorter than healthy plants and may sometimes be half to that of healthy plants. Plants infected with the dwarf bunt fungus may be only one fourth as tall as healthy plants and may show an increase in number of tillers. Infected plants may appear slightly bluish green to greyish green in colour, but this is not easily distinguishable.

Distinct bunt symptoms, however, are shown by the heads of infected plants. Infected heads are slimmer and are usually bluish green rather than the normal yellowish green, and their glumes seem to spread apart and form a greater angle with the main axis. Infected kernels are shorter and thicker than healthy ones and are greyish brown rather than the normal golden yellow or red. When mature kernels are broken, they are found to be full of a sooty, black, powdery mass of fungus spores that give off a distinctive odour resembling that of decaying fish. The bunt infected field emit a fishy odour. During the harvest of infected fields, large clouds of spores may be released in the air.

Fig. 142: Bunt disease symptoms and its pathogen

3.2.6. Leaf Spot disease causing fungi

Alternaria, Helminthosporium, Diplodia, Pyricularia, Cercospora, Septoria, Physoderma, Cladosporium, Colletotrichum, Mycosphaerella, Magnaporthe, Cochliobolus, Botryosphaeria, Diaporthe, Gnomonia, Phyllosticta, Taphrina, Ascochyta, Marssonia, Rhytisma etc.

The fungal pathogens infect several important crop plant species (Fig.143) to cause leaf spot disease

Most leaf spot diseases develop as small, scattered, circular to oval dead areas in the leaves; usually tan, dark brown, yellow, gray, purple, or black. Some spots are raised, shiny, and coal black, others may drop out leaving ragged holes; some are marked with light and dark concentric zones. Numerous spots develop yellow, purple, red, or reddish brown to black margins; and later, in damp weather, increase in size and number and merge into large, angular to irregular dead areas. Dark areas and speck-sized fungal fruiting bodies (known as pycnidia, acervuli, and perithecia) commonly form in the dead tissues of many older spots. Heavily infected leaves may turn yellow to brown, wither, and drop early, weakening the tree. Occasionally, some leaf spotting fungi deform or kill flowers, buds, fruits, twigs, or even small branches.

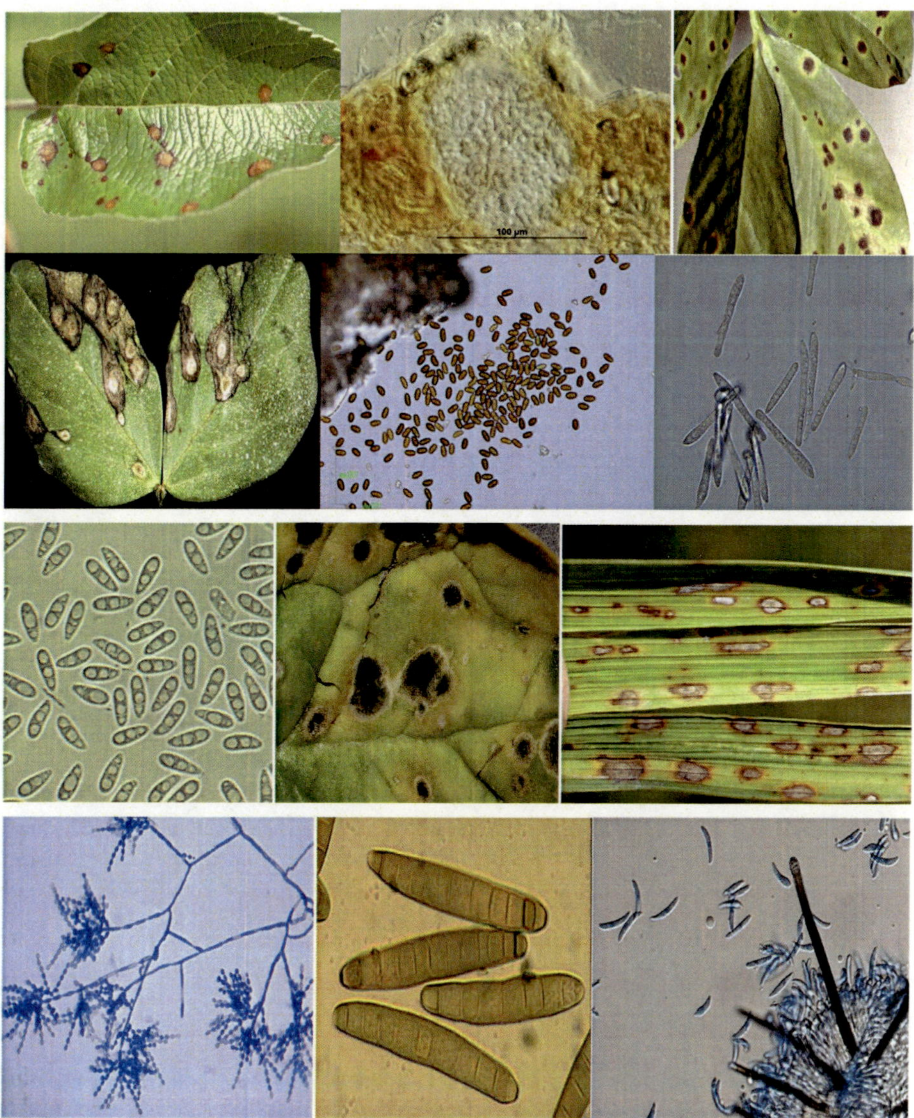
100 µm

Fig. 143: Leaf spot disease symptoms and its pathogens

3.2.7. Blight disease causing fungi

Alternaria, Helminthosporium, Colletotrichum, Phytophthora, Cochliobolus, Setosphaeria, Cryphonectria, Monilinia, Didymella etc.

The fungal pathogen infects several import crop plant species to cause blight disease (Fig.144).

Blights are the general and rapid destruction of the growing succulent tissues like leaves, shoots, twigs and blossoms covering the large area within small period of infection. The blighted tissue often gives the appearance of tissue being burnt with fire.

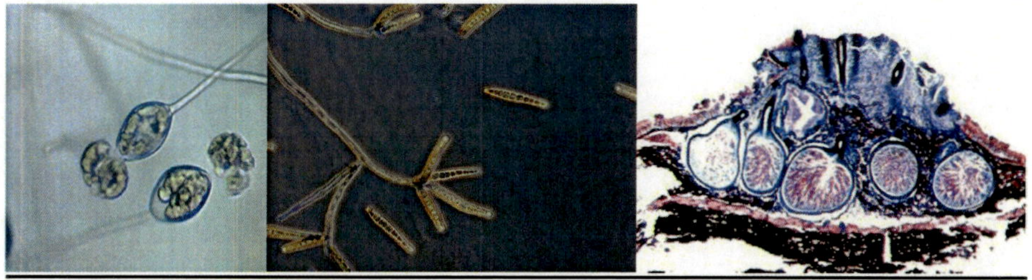

Fig. 144: Blight disease symptoms and its pathogens

3.2.8. Canker disease causing fungi

Nectria, Leucostoma, Diaporthe, Phomopsis, Cytospora, Hypoxylon *etc*

The fungus pathogen infects *Apple, pear and other ornamental plants.*

The symptoms of Cankers are usually oval to elongate, localized, sunken, slightly discoloured, brown-to-reddish lesions on the bark of trunks and branches, or as injured areas on smaller twigs but can vary considerably in size and shape (Fig.145). The bark often splits between the diseased and the healthy tissue, and sometimes it may ooze sap or moisture. The inner bark turns black and sometimes gives off a foul odour. The newest leaves on affected branches are usually the first to show decline symptoms. Leaves may appear smaller than normal, pale green to yellow or brown, often curled and sparse. As the fungal pathogen invades bark and sapwood, the water-conducting tissues (vascular system) become blocked or dies, causing wilting and dieback to occur. Cankers are formed by the interaction between the host and pathogen. The pathogen grows within the wood and the host tree tries to contain the growth. Cankers can take months (or years) to enlarge enough to girdle twigs, branches, or trunks.

Fig. 145: Canker disease symptoms and its pathogens

3.2.9. Scab disease causing fungi

Venturia, Elsinoe, Sphaceloma, Cladosporium etc

The fungal pathogen infects Apple, Citrus, Avocado, Cucumber to cause scab disease.

The disease symptoms are characterised by the localized, slightly raised and cracked lesions formed on the outer layer of the fruits skin, leaves or tubers etc where the cracked tissues become dry and corky (Fig.146). The fungal disease forms pale yellow or olive-green spots on the upper surface of leaves. Dark, velvety spots may appear on the lower surface. Severely infected leaves become twisted and puckered and may drop early in the summer.

Symptoms on fruit are similar to those found on leaves. Scabby spots are sunken and tan and may have velvety spores in the centre. As these spots mature, they become larger and turn brown and corky. Infected fruit becomes distorted and may crack, allowing entry of secondary organisms. Severely affected fruit may drop, especially when young.

Fig. 146. Scab disease symptoms and its pathogens

3.2.10. Ergot disease causing fungi

***Claviceps* sp.**

The fungal pathogen infects Cereals to cause ergot disease.

The first symptoms appear as creamy droplets of a sticky liquid exuding from young florets of infected heads. The droplets are soon replaced by a hard, hornshaped, purplish-black fungal mass a few millimeters in diameter and 0.2 to 5.0 centimeters long (Fig.147). These are the sclerotia or ergots of the fungus that grow in place of the kernel and consist of a hard compact mass of fungal tissues.

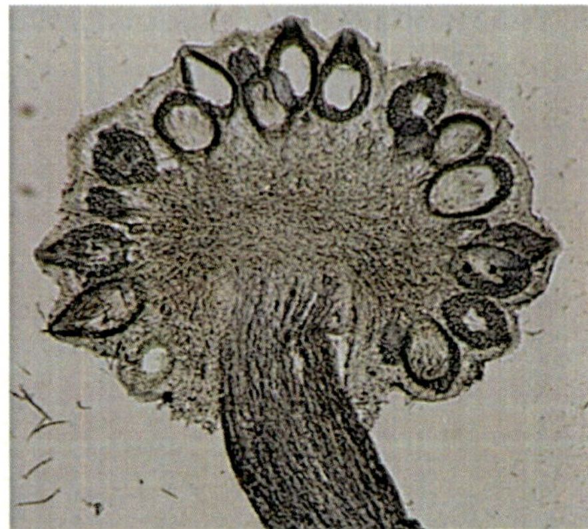

Fig. 147: Ergot disease symptoms and its pathogens

3.2.11. Anthracnose disease causing fungi

Colletotrichum, Elsinoe, Diplocarpon

The fungal pathogen infects Allium crops plants such as onion, garlic, leek., solanaceous plant species such as pepper, tomato, chilli , potato., beans, cucurbits, lettuce, cereals, cassava, cotton, coffee, strawberry etc.

The Anthracnose disease pathogen attacks all plant parts at any growth stage. The symptoms are most visible on leaves and ripe fruits. At first, anthracnose generally appears on leaves as small and irregular yellow, brown, dark brown, or black spots. The spots can expand and merge to cover the whole affected area. The colour of the infected part darkens as it ages. The disease can also produce cankers on petioles and on stems that causes severe defoliation and rotting of fruits and roots. Infected fruit has small, water soaked, sunken, circular spots that may increase in size up to 1.2 cm in diameter. As it ages, the centre of an older spot becomes depressed greyish with blackish margin (Fig.148) and emits gelatinous pink spore masses.

Fig. 148: Anthrocnose disease symptoms and its pathogens

3.3 Fungi involved in post harvest losses

3.3.1. Post harvest losses of fruits and vegetables

The most common fungi responsible for post harvest losses of fruits and vegetables are ***Aspergillus, Penicillium, Rhizopus, and Mucor.*** All of them are found commonly to cause moulding of bread, whereas ***Penicillium*** and ***Rhizopus*** cause post-harvest rots of numerous kinds of wounded or senescent fruits and vegetables. ***Aspergillus*** is found more commonly causing moulding of grains and legumes; while ***Rhizopus*** causes many fruit rots, as in peach and strawberry, whereas ***Penicillium*** causes the rotting of many wounded fruit, e.g., pears.

3.3.1.1. Post harvest Spoilage due to alternaria

The species of ***Alternaria*** fungus causes post harvest spoilage of several Species of Fruits & Vegetables (Fig.149).

The various species of ***Alternaria*** cause decay on most, if not all, fresh fruits and vegetables either before or after harvest. Symptoms appear as brown or black, flat or sunken spots with definite margins, or as diffuse, large, decayed areas that are shallow or extend deep into the flesh of the fruit or vegetable. The fungus develops well at a wide range of temperatures, even in the refrigerator, although at a slower rate. The fungus may spread into and rot tissues internally with little or no mycelium appearing on the surface, but usually a mat of mycelium that is white at first but later turns brown to black forms on the surface of the rotted area.

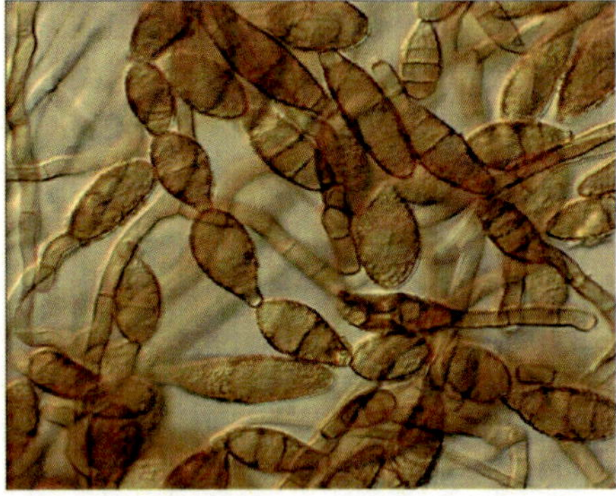

Fig. 149: Alternaria post harvest disease symptoms and its pathogen

3.3.1.2. Post harvest fpoilage gue to Botrytis

The species of *Botrytis* fungus infects fruits and vegetables of several species to cause post harvest spoilage (Fig.150).

Botrytis causes the grey moulds or grey mould rots of fruits and vegetables, both in the field and in storage. Almost all fresh fruits, vegetables, and bulbs are attacked by ***Botrytis*** in storage. Some products, such as strawberry, lettuce, onion, grape, and apple, are also attacked in the field near maturity or while green.

The decay may start at the blossom or stem end of the fruit or at any wound. The decay appears as a well-defined water-soaked, then brownish, area that penetrates deeply and advances rapidly into the tissue. In most hosts and under humid conditions a greyish or brownish-grey, granular, mould layer develops on the surface of decaying areas. Grey moulds are most severe in cool, humid environments and continue to develop, although slowly, even at 0°C.

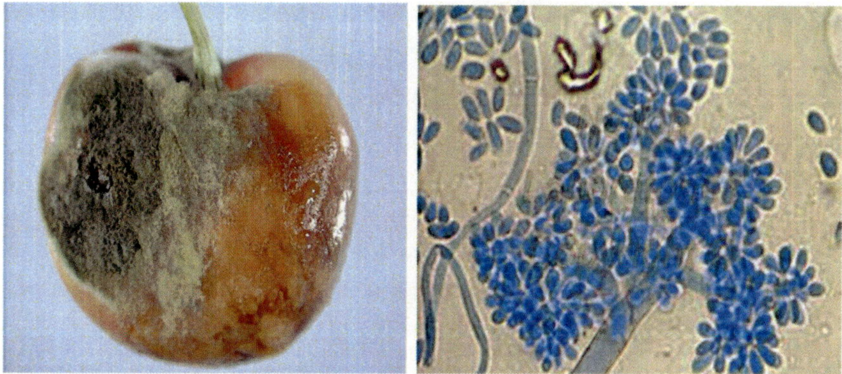

Fig. 150: Botrytis post harvest disease symptoms and its pathogen

3.3.1.3. Post harvest spoilage due to Fusarium

The species of ***Fusarium*** fungi infects fruits and vegetables of several species to cause post harvest spoilage (Fig.151).

Fusarium develops into post-harvest pink or yellow moulds on vegetables, ornamentals and especially on root crops, tubers, and bulbs. Low-lying crops such as cucurbits and tomatoes are also affected frequently. Contamination with *Fusarium* usually takes place in the field before or during harvest, but infection may develop in the field or in storage. Losses are particularly heavy with crops such as potatoes that are stored for long periods of time.

Affected tissues appear fairly moist and light brown at first, but later they become darker brown and somewhat dry. As the decaying areas enlarge, they often become sunken, the skin is wrinkled, and small tufts of whitish, pink, or yellow mould appear. The infection of softer tissues such as tomatoes and cucurbits develops faster and is characterized by pink mycelium and pink, rotten tissues.

Fig. 151: Fusarium post harvest disease symptoms and its pathogen

3.3.1.4. Post harvest spoilage due to Geotrichum

The species of *Geotrichum* fungal pathogen infects different species of Fruits and Vegetables to cause post harvest spoilage (Fig.152).

Geotrichum causes the sour rots of citrus fruits, tomatoes, carrots, and other fruits and vegetables. Sour rot is one of the messiest and most unpleasant rots of susceptible fruits and vegetables. Although it may affect them at the mature green stage, it is the ripe or over-ripe fruits and vegetables and those kept in moisture-holding plastic bags or packages that are particularly susceptible to sour rot.

The fungus occurs in soils, decaying fruits and vegetables and contaminates new ones before or during harvest. The fungus penetrates fruits, usually after harvest, at wounds of various sorts.

Infected areas appear water soaked and soft and are punctured easily. The decay spreads rapidly. Later, the skin frequently cracks over the affected area and is usually filled with a white, cheesy, or scum-like development of the fungus. Also, a thin, water-soaked layer of compact, cream-colored fungal growth develops on the surface, while the whole inside becomes a sour smelling, decayed, watery mass. Fruit flies, which are attracted to tissues affected with sour rot, spread the pathogen further. The fungus prefers high temperatures (24–30°C) and humidity but is active at temperatures as low as 2°C.

Fig. 152: Geotrichum post harvest disease symptoms and its pathogen

3.3.1.5. Post harvest Spoilage due to Penicillium

The species of *Penicillium* fungal pathogen infects Fruits and Vegetables of several Species to cause post harvest spoilage (Fig.153).

Various species of *Penicillium* cause the blue mould rots and green mould rots, known as *Penicillium* rots. They are the most common and usually the most destructive of all post harvest diseases, affecting most kinds of fruits and vegetables. On some fruits, such as citrus, some infections may take place in the field, but blue moulds or green moulds are essentially post harvest diseases and often account for up to 90% of decay in transit, in storage, and in the market. *Penicillium* enters tissues through wounds. However, it can spread from infected fruit in contact with healthy ones through the uninjured skin.

Penicillium rots at first appear as soft, watery, slightly discoloured spots of varying size and on any part of the fruit. The spots are rather shallow at first but quickly become deeper. At room temperature most or all of the fruit decays in just a few days. Soon a white mould begins to grow on the surface of the fruit, near the centre of the spot, and starts producing spores. The sporulating area has a blue, bluish-green, or olive-green colour and is usually surrounded by white mycelium and a band of water-soaked tissue. The fungus develops on spots of any size as long as the air is moist and warm. In cool, dry air, surface mould is rare, even when the fruits are totally decayed. Decaying fruit has a musty odour. Under dry conditions it may shrink and become mummified. Under moist conditions, secondary fungi and yeasts also enter the fruit, which is then reduced to a wet, soft mass.

In addition to the losses caused by the rotting of fruits and vegetables by *Penicillium*, the fungus also produces several mycotoxins, such as patulin, in the affected products, which contaminate juices and sauces made from healthy and partly rotten fruits. Such mycotoxin contaminated products are injurious to health.

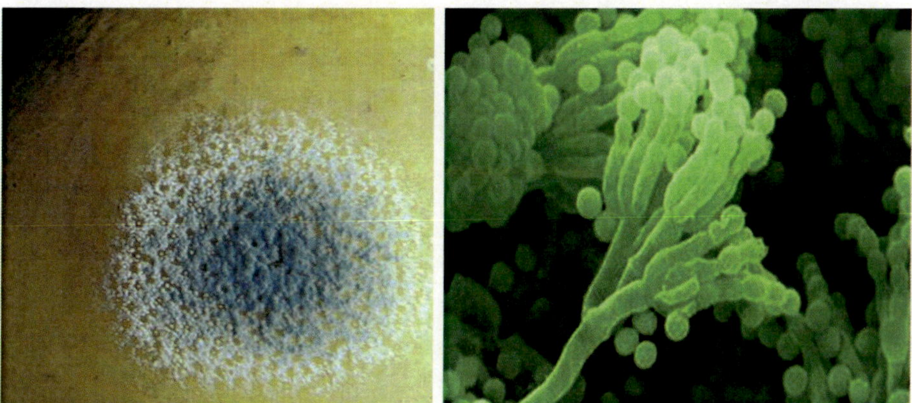

Fig. 153: Penicillium post harvest disease symptoms and its pathogen

3.3.1.6. Post harvest fpoilage gue to Sclerotinia

The species of *sclerotinia* fungal pathogen infects many Fruit and Vegetable species to cause post harvest spoilage (Fig.154). *Sclerotinia* causes the cottony rot of citrus fruits, especially lemons, and the watery soft rot of many fruits and practically all vegetables except onions and potatoes.

In a moist atmosphere, a soft, watery decay is produced, and the affected tissues are covered rapidly with a white, cottony growth of mycelium that is characteristic of this decay. In moist air, succulent decaying products may be completely liquefied, leak, and leave a pool of juice. In dry air the water may evaporate as fast as it is liberated by the decay, and the tissues dry down into a mummy or parchment-like remains. Cottony rot is a rapidly spreading, contact decay that attacks both green and mature fruits and vegetables.

Black sclerotia, 2 to 15 mm long, later develop in the fungus mat. The activity of the fungus and the severity of the rot increase with temperature up to 25°C, but, once started, rotting of tissues continue at temperatures as low as 0°C.

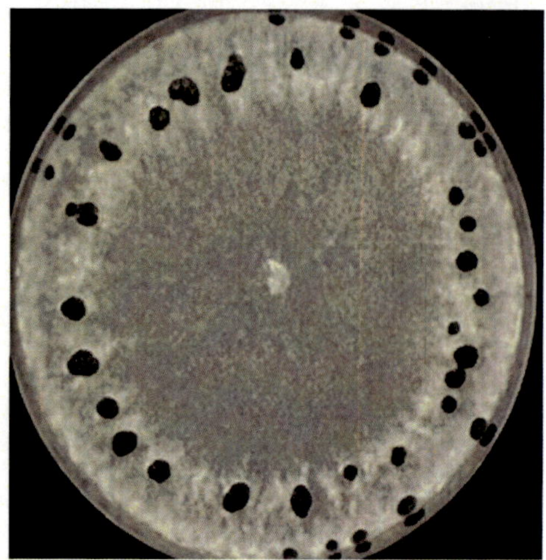

Fig. 154: *Sclerotinia* post harvest disease symptoms and its pathogen

3.3.1.7. Post harvest fpoilage due to Rhizopus

The *Rhizopus* Fungus causes soft rot of several species of Fruits and Vegetables during post harvest stages (Fig.155).

Infected fleshy parts appear water soaked at first and are very soft. If the skin of the infected parts remains intact, the tissue loses moisture gradually until it shrivels into a mummy. More frequently, however, fungal hyphae grow outward through the wounds and cover the affected portions by producing tufts of whisker-like grey sporangiophores and sporangia. The bushy growth of the fungus often spreads over the surface of the healthy portions of affected fruit and even to the surface of the containers when they become wet with the exuding liquid. Affected tissues at first give off a mildly pleasant smell, but soon yeasts and bacteria move in and a sour odour develops. When loss of moisture is rapid, infected organs finally dry up and mummify. If the loss of moisture is slow, they break down and disintegrate in a "leaky" watery rot.

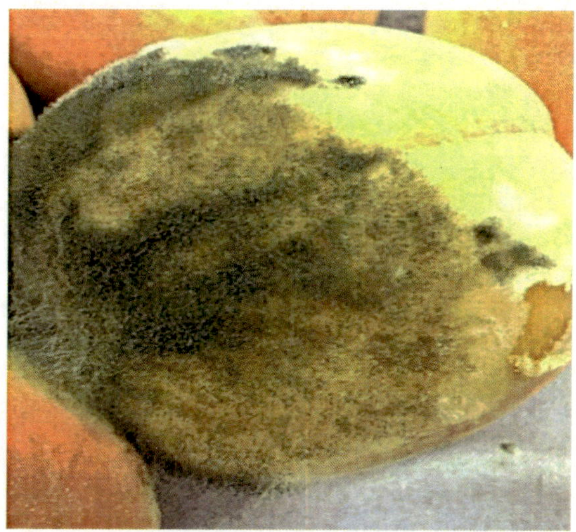

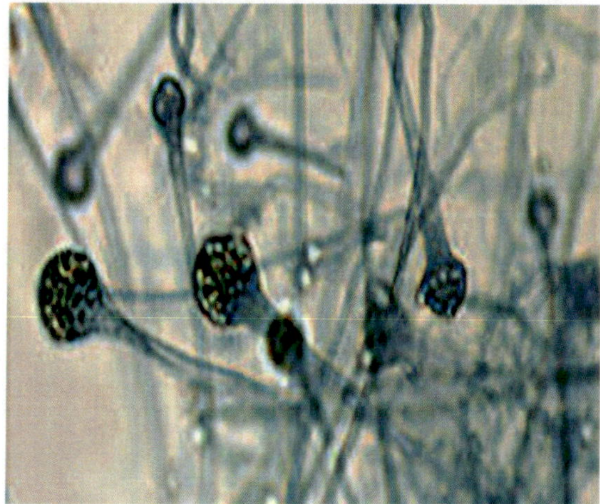

Fig. 155: *Rhizopus* post harvest disease symptoms and its pathogen

3.3.2. Post harvest Decays of Grain and Legume Seeds

The fungus infecting grains and legume seeds not only grow on them to destroy their quality (Fig.156) but also produce the toxins in the infected seeds which causes mycotoxicosis and health hazards in human beings, animals & birds which consume such seeds. In severe cases death is ultimately envitable. The fungi which infect and produce toxins during post harvest decays are species of ***Aspergillus.***

3.3.2.1. Aspergillus

The ***Aspergillus*** fungus produce toxin known as *Aflatoxins.* Aflatoxins are produced by *Aspergillus flavus* and several other species of *Aspergillus*. Aflatoxins are produced in infected cereal seeds and most legumes, but they often reach a rather low and probably nontoxic concentration (about 50 ppb). During some years, a rather high percentage (30% or more) of the corn harvest over large areas contains more than 100 ppb aflatoxin, which is five times that allowed in food for humans and in feed for sensitive animals such as chickens. However, in peanuts, cottonseed, fishmeal, Brazil nuts, and probably other seeds or nuts grown in warm and humid regions, aflatoxin is produced at high concentrations (up to 1000 ppb or more) and causes mostly chronic or occasionally acute mycotoxicoses in humans and domestic animals. Aflatoxins exist in a variety of derivatives with varying effects. Some of these toxins, when ingested with the feed by dairy cattle, are excreted in the milk in still toxic form.

Fig.156:. *Aspergillus* post harvest decays and its pathogen

Symptoms of *Aflatoxin* Poisoning

The symptoms of mycotoxicoses caused by aflatoxin in animals, and presumably in humans, vary widely with the particular toxin and animal species, dosage, age of the animal, and so on. Young ducklings and turkeys fed with high dosages of aflatoxin become severely ill and die. Pregnant cows, calves, fattening pigs, mature cattle, and sheep fed with low dosages of aflatoxin over long periods develop weakening, intestinal bleeding, debilitation, reduced growth, nausea, refusal of feed, predisposition to other infectious diseases, and may abort. Moreover, most of the ingested aflatoxin is taken up by the liver, and, in some experiments, animals given feed containing even less than the permissible amount of aflatoxin (20 ppb) almost invariably developed liver cancer.

3.3.2.2. Fusarium

The fungus Fusarium produce three groups of toxins i.e. **zearalenones**, **trichothecenes**, and **fumonisins**, primarily in mouldy corn (Fig.157).

Zearalenones seem to be most toxic to swine, in which they cause abnormalities and degeneration of the reproductive system, the so-called estrogenic syndrome. Female swine fed with zearalenone containing feed develop swollen vulvas bearing bleeding lesions and atrophying, non-functioning ovaries. They are susceptible to abortion, and piglets that are born are small and weak. Male swine show signs of feminization, namely atrophy of the testes and enlargement of the mammary glands.

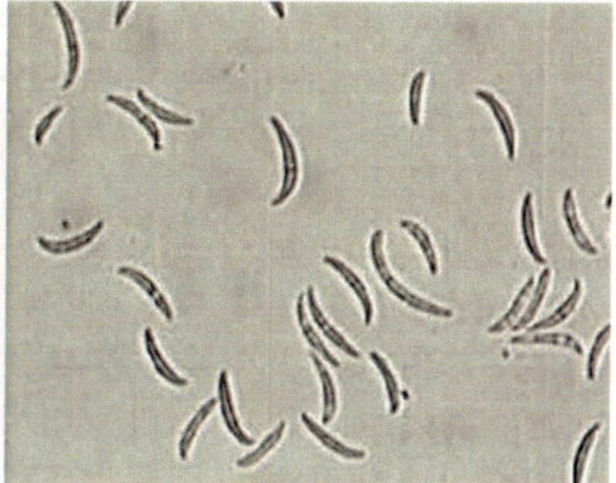

Fig.157: *Fusarium* post harvest decays and its pathogen

Fumonisins are produced by ***Fusarium moniliforme***, which causes Fusarium ear rot of corn that affects as much as 90% of the corn fields. Fumonisins are the cause of blind staggers (equine leukoencephalomalacia) in horses, donkeys and mules, pulmonary edema in swine, and, possibly, cancer in humans.

Trichothecins or **trichothecenes**, are most toxic when fed to swine, in which they cause, among other symptoms, listlessness or inactivity, degeneration of the cells of the bone marrow, lymph nodes, and intestines, diarrhea, bleeding, and death. Other animals, however, such as cows, chicks, and lambs, are also affected.

Other Symptoms of the toxin on various body part are:

– Eye pain, excessive lacrimation, visual blurring and scleral injection.

– Nasal itching and pain, epistaxis, cough, dyspnea, wheezing.

– Burning skin, redness, blistering progressing to necrosis, skin sloughing.

l – Anorexia, mouth pain, nausea, vomiting, hematemesis, abdominal pain, watery or bloody diarrhea, abdominal cramps.

- *Early systemic effects* – Weakness, loss of coordination, dizziness, ataxia, tachycardia, hypothermia, hypotension or death.
- *Late systemic effects* – Two to eight weeks after ingestion on contaminated food, bone marrow suppression occurs with severe neutropenia and hemorrhagic syndromes such as diffuse bleeding into skin with petechiae, melena, hematuria, hematemesis, epistaxis, and vaginal bleeding. Other common problems include fever, oral and GI ulceration, and secondary sepsis. These effects are similar to the effects seen with exposure to radiation.

Other toxin produced by Fusarium species is **Deoxynivalenol**, also known as **vomitoxin** or **DON** is produced by the fungus *Gibberella zeae* (anamorph *Fusarium graminearum*), the cause of Gibberella ear rot of corn and of head blight (scab) of wheat. The mycotoxin at first causes reduced feeding by the animals and, thereby, slower gain or loss of weight. At higher concentrations of the mycotoxin, the animals are induced to vomit and totally refuse to eat.

3.3.2.3. Penicillium

The fungus Penicillium (Fig.158) produce the toxin Patulin. It is commonly found to occur naturally in foodstuffs such as fruit or juices made with fruit partly infected with *Penicillium*, in naturally moulded bread and bakery products, and in most commercial apple products.

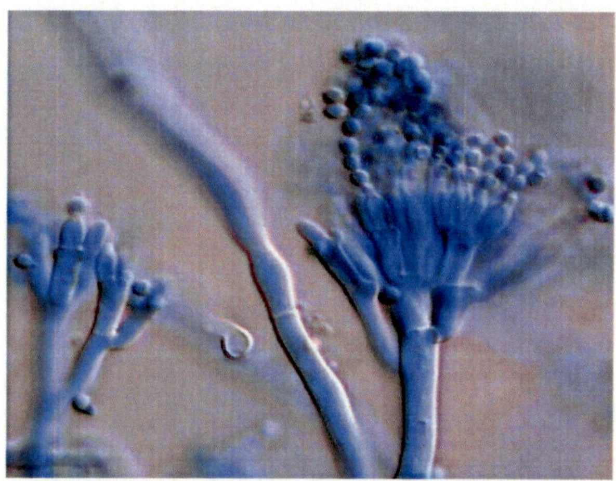

Fig.158: *Penicillium* as post harvest pathogen

Thus, patulin may constitute a serious health hazard for humans as well as for animals. *It causes edema and bleeding in lungs and brain*

Symptoms induced by Patulin:

- Damage to kidneys,
- Paralysis of motor nerves and it also induces cancer in higher organisms.

2.3.2.3.1. Yellow Rice Toxins

These toxins are primarily citreoviridin, citrinin, and luteoskyrin, and are all produced by species of *Penicillium* growing in stored rice, barley, corn, and dried fish. They cause toxicoses associated with various diseases, nervous and circulatory disorders, and degeneration of the kidneys and liver

2.3.2.3.2. Ochratoxin

This mycotoxin is primarily produced by species of *Penicillium* and *Aspergillus*. Ochratoxin is primarily a **kidney toxin** but if the concentration is sufficiently high there can be damage to the liver as well. It occurs in wide variety of commodities such as raisins, barley, soy products and coffee.

The symptoms produced by this toxin includes Anemia, Anorexia (Refusal to eat), Apoptosis (cell death), Carcinogenic, Copper coloured skin, decreased lymphocytes (immune cells), Endemic Nephropathy, Fatigue, headache, increased apoptotic phagocytes (death of neutrophils and macrophages), Increased clotting time, increased reactive oxygen radicals (very powerful oxidizing agents that cause structural damage to proteins and nucleic acids)

3.3.2.4. Claviceps

The *sclerotia* (ergots) of the ergot fungus *Claviceps (Fig.159) contains the toxic substances* and its effect is the oldest known *mycotoxicosis* known as *ergotism*.

The *mycotoxicosis* of *claviceps* causes:

- Convulsions and limb swellings
- Gangrene of body extremities
- Burning sensations

Fig.159: *Fungal Sclerotia* as post harvest contaminant.

3.3.2.5. Acremonium

The *Acremonium* toxin is produced on plants of the perennial tall fescue grass infected systemically with the fungus *Acremonium (Fig.160)*. The fungus is an endophyte growing internally through the plant without invading its cells. The fungus actually seems to make the infected plants more resistant to stress, particularly drought.

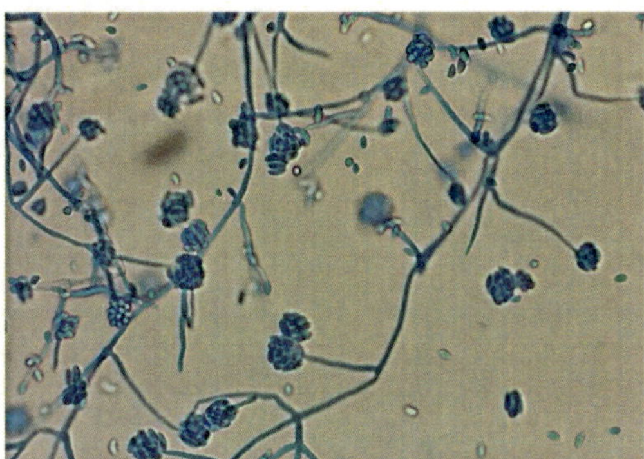

Fig.160: Fungal spores of *Acremonium*.

Horses eating tall fescue plants infected with the fungus show only reproductive disorders. Cattle feeding on such plants, in addition to reduced calving and lower milk production, show reduced weight gains, elevated body temperature, and rough hair coat; moreover as in ergotism, feet or other body extremities may develop gangrene and drop off ("fescue foot").

4

Fungi Causing Diseases in Human Beings

4.1. *Aspergillosis* disease causing fungi

Aspergillus

Aspergillosis is usually caused by the fungus *Aspergillus* (Fig.161) and usually occurs in people with lung diseases or weakened immune system

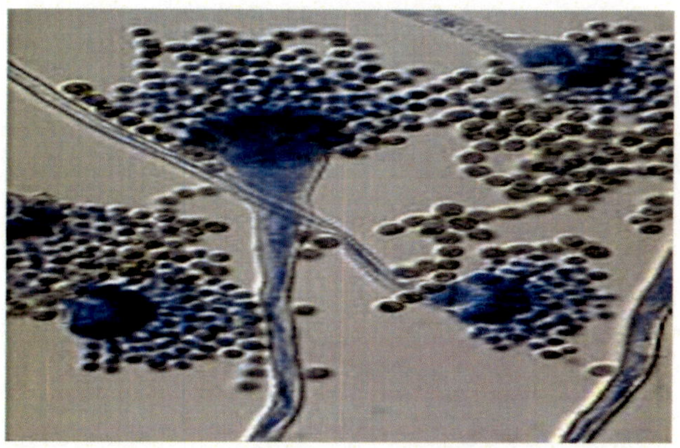

Fig. 161: Fungal structure and spores of *Aspergillus*

The different types of aspergillosis can cause different symptoms. The symptoms are similar to asthma symptoms, and include:

4.1.1. Allergic bronchopulmonary aspergillosis (ABPA)

The symptoms of ABPA includes Wheezing, Shortness of breath, Cough and Fever (in rare cases).

4.1.2. Allergic Aspergillus sinusitis

The symptoms of Alleric Aspergillus Sinusitis include stuffiness, Runny nose, Headache and Reduced ability to smell.

4.1.3. Aspergilloma ("fungus ball")

The symptoms of Aspergilloma include Cough, Coughing up blood and Shortness of breath.

4.1.4. Chronic pulmonary aspergillosis

The symptoms of chronic pulmonary aspergillosis include Weight loss, Cough, Coughing up blood, Fatigue and Shortness of breath

4.1.5. Invasive Aspergillosis

It usually occurs in people who are already sick from other medical conditions, so it can be difficult to know which symptoms are related to an *Aspergillus* infection. Fever is a common symptom of invasive aspergillosis. However, the symptoms of invasive aspergillosis in the lungs include Fever, Chest pain, Cough, Coughing up blood, Shortness of breath.

Other symptoms can develop if the infection spreads from the lungs to other parts of the body.

4.2. *Blastomycosis* disease causing fungi

Blastomyces dermatitidis

Blastomycosis is a disease caused by the fungus *Blastomyces dermatitidis (Fig.162)*. The fungus lives in moist soil and in association with decomposing organic matter such as wood and leaves. Lung infection can occur after a person inhales airborne, microscopic fungal spores from the environment; however, many people who inhale the spores do not get sick. The symptoms of blastomycosis are similar to flu symptoms, and the infection can sometimes become serious if it is not treated

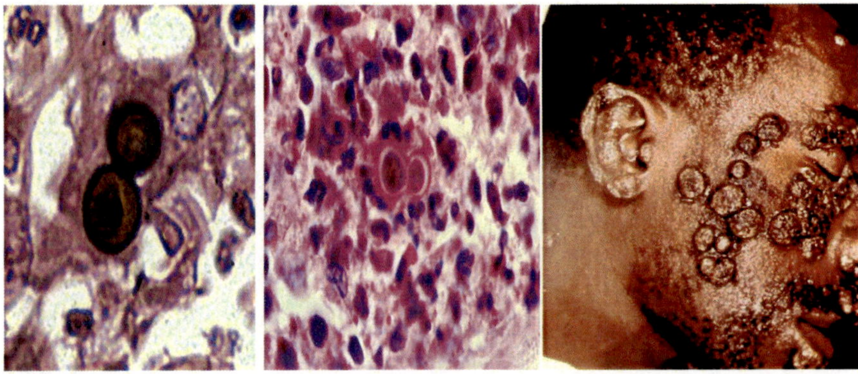

Fig. 162: Fungal structures with spores of *Blastomyces dermatitidis* and it symptoms

Only with about half of the people who are infected with blastomycosis will show symptoms. If symptoms occur, they usually appear between 3 and 15 weeks after being exposed to the fungus.

The symptoms of blastomycosis are similar to flu symptoms, and include fever and night sweats, chills, cough may produce brown or bloody mucus, muscle aches, bone and joint pain, and chest pain, Fatigue, General discomfort, uneasiness, or ill feeling (malaise), and unintentional weight loss. In very serious cases of blastomycosis, the fungus can disseminate (spread) to other parts of the body, such as the skin and bones.

4.3. *Candidiasis* disease causing fungi

Candida spp

Candidiasis is a fungal infection caused by yeasts that belong to the genus *Candida* (Fig.163). There are over 20 species of *Candida* yeasts that can cause infection in humans, the most common of which is *Candida albicans*.

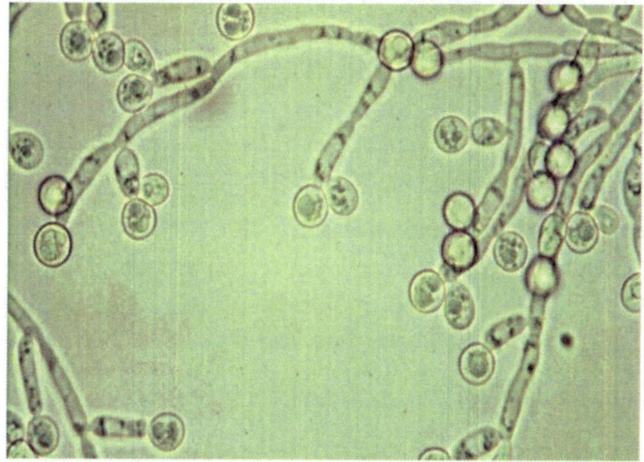

Fig. 163: Fungal structures with spores of *Candida* spp.

Candida yeasts normally live on the skin and mucous membranes without causing infection; however, overgrowth of these organisms can cause symptoms which vary depending on the area of the body that is infected.

Candidiasis that develops in the mouth or throat is called thrush or oropharyngeal candidiasis. Candidiasis in the vagina is commonly referred to as a "yeast infection". Invasive candidiasis occurs when *Candida* species enter the bloodstream and spread throughout the body.

4.3.1. Thrush or Oropharyngeal / Esophageal Candidiasis

Candidiasis that develops in the mouth or throat is called "thrush" or oropharyngeal candidiasis. *Candida* infections of the mouth and throat can manifest in a variety of ways. The most common symptom of oral thrush is white patches or plaques on the tongue and other oral mucous membranes (Fig.164).

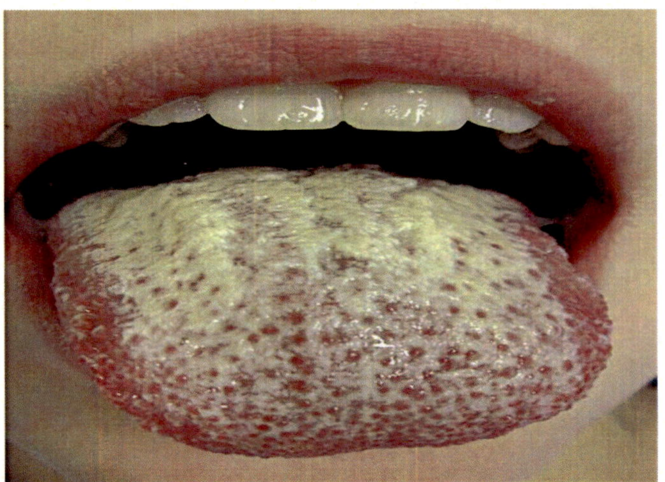

Fig. 164: Infection of *Candida* sp in mouth as oropharyngeal candidiasis

Other symptoms include redness or soreness in the affected areas, difficulty in swallowing and Cracking at the corners of the mouth (angular cheilitis)

4.3.2. Vaginal Candidiasis ("Genital / Vulvovaginal Yeast Infection")

Genital / vulvovaginal candidiasis (VVC) is a "yeast infection," and it occurs when there is overgrowth of the yeast in the vagina (Fig.165).

Women with VVC usually experience genital itching, burning, and sometimes a "cottage cheese-like" vaginal discharge. Men with genital candidiasis may experience an itchy rash on the penis. The symptoms of VVC are similar to those of many other genital infections, so it is important to see your doctor if you have any of these symptoms.

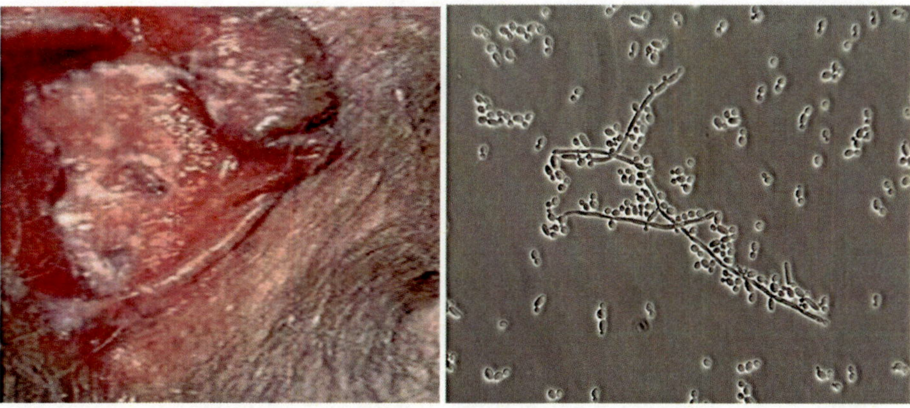

Fig. 165: Infection of *Candida* sp as VCC

This infection is relatively common i.e nearly 75% of all adult women have had at least one "yeast infection" in their lifetime.

4.3.3. Invasive Candidiasis

Invasive candidiasis is an infection caused by a yeast called *Candida.* Unlike *Candida* infections in the mouth and throat or vaginal infections, invasive candidiasis is a serious infection that can affect the blood, heart, brain, eyes, bones, and other parts of the body. Candidemia, a bloodstream infection with *Candida (Fig.166)*, is a common infection in hospitalized patients.

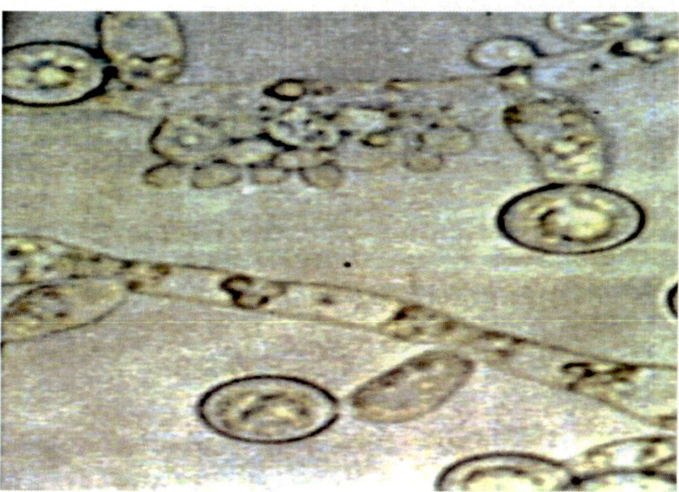

Fig. 166: *Candida* sp in Invasive candidiasis

People who develop invasive candidiasis are often already sick from other medical conditions, so it can be difficult to know which symptoms are related to a *Candida* infection. However, the most common symptoms of invasive candidiasis are fever and chills that don't improve after antibiotic treatment for suspected bacterial infections. Other symptoms can develop if the infection spreads to other parts of the body, such as the heart, brain, eyes, bones, or joints.

4.4. Valley Fever (Coccidioidomycosis) disease causing fungi

Coccidioides

Valley fever, also called coccidioidomycosis, is an infection caused by the fungus *Coccidioides* (Fig.167). People can get valley fever by breathing the microscopic fungal spores from the air, although most people who breathe in the spores don't get sick. Usually, people who get sick with valley fever will get better on their own within weeks to months, but some people will need antifungal medication. Certain groups of people are at higher risk for becoming severely ill. It's difficult to prevent exposure to *Coccidioides* in areas where it is common in the environment, but people who are at higher risk for severe valley fever should try to avoid breathing in large amounts of dust if they are in these areas.

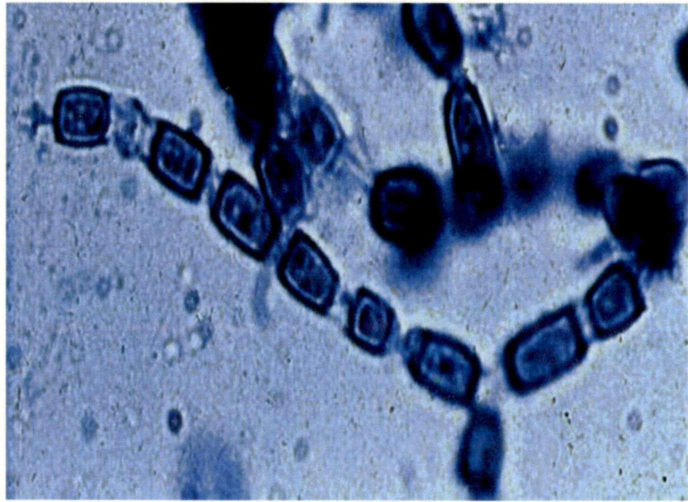

Fig. 167: Spores of *Coccidioides* fungus

The symptoms of valley fever include fatigue (tiredness), Cough, Fever, Shortness of breath, Headache, Night sweats, Muscle aches or joint pain and Rash on upper body or legs (Fig.168).

Fig. 168: Symptoms of rashes on the body by infection of *Coccidioides* fungus

In extremely rare cases, the fungal spores can enter the skin through a cut, wound, or splinter and cause a skin infection

The fungus is known to live in the soil in the southwestern United States and parts of Mexico and Central and South America. The fungus was also recently found in south-central Washington.

4.5. *Cryptococcus neoformans* fungal Infection

Cryptococcus neoformans is a yeast fungus (Fig.169) that lives in the environment throughout the world. People can become infected with *C. neoformans* after breathing the yeast fungus, although most people who are exposed to the fungus never get sick from it. *C. neoformans* infections are extremely rare in people who are otherwise healthy. Most cases occur in people who have weakened immune systems, particularly those who have advanced HIV/AIDS.

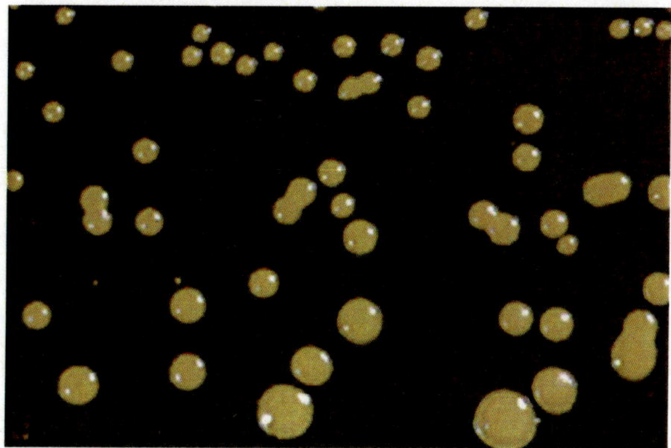

Fig. 169: Yeast colonies of *Cryptococcus neoformans* on growth medium

C.neoformans usually infects the lungs or the central nervous system (the brain and spinal cord), but it can also affect other parts of the body. The symptoms of the infection depend on the parts of the body that are affected (170).

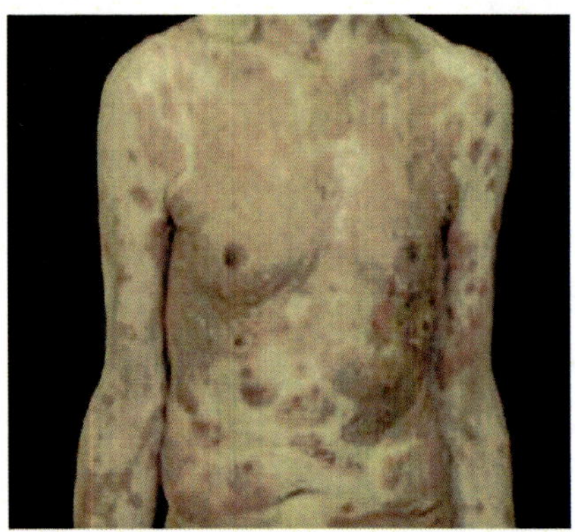

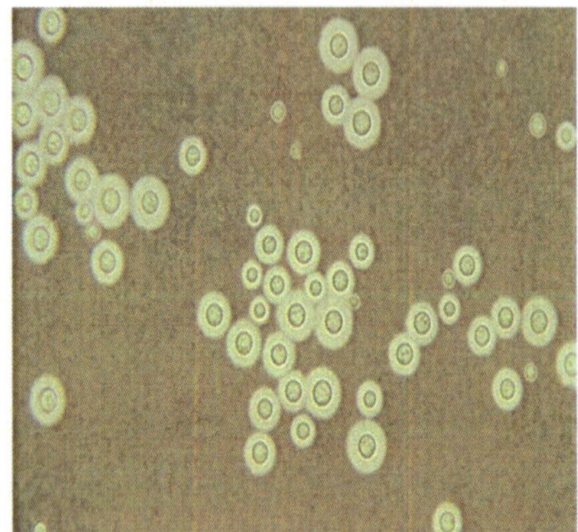

Fig. 170: Symptoms of *Cryptococcus neoformans* infection on body and the pathogenic yeast cells

C.neoformans infection in the lungs (Fig.171) can cause a pneumonia-like illness. The symptoms are often similar to those of many other illnesses, and can include Cough, Shortness of breath, Chest pain, Fever.

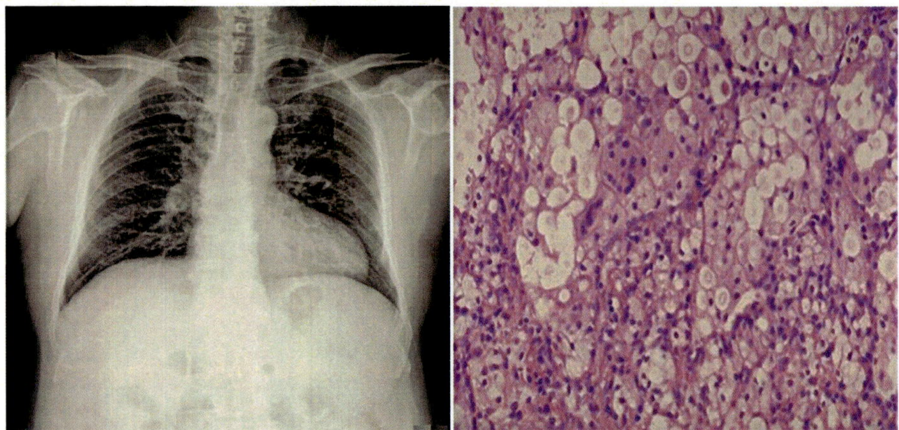

Fig. 171: Symptoms of *Cryptococcus neoformans* infection in lung and the pathogenic yeast cells

The infection of Cryptococcal neoformans can spreads from the lungs to the brain and this infection is known as Cryptococcal meningitis. The symptoms of cryptococcal meningitis include Headache, Fever, Neck pain, Nausea and vomiting, Sensitivity to light, Confusion or changes in behavior

4.6. *Cryptococcus Gattii* fungal Infection

Cryptococcus gattii is a yeast fungus that lives in the environment in many tropical and sub-tropical areas of the world as well as British Columbia and the U.S. Pacific Northwest. *C. gattii* cryptococcosis is a rare infection that people can get after breathing in the microscopic fungus.

C.gattii usually infects the lungs or the central nervous system (the brain and spinal cord), but it can also affect other parts of the body. The symptoms of the infection depend on the parts of the body that are affected.

C. gattii infection in the lungs can cause a pneumonia-like illness. The symptoms are often similar to those of many other illnesses, and can include Cough, Shortness of breath, Chest pain and Fever

C.gattii infection can also cause cryptococcomas (fungal growths) to develop in the lungs, skin, brain or other organs, causing symptoms in other parts of the body (Fig.172).

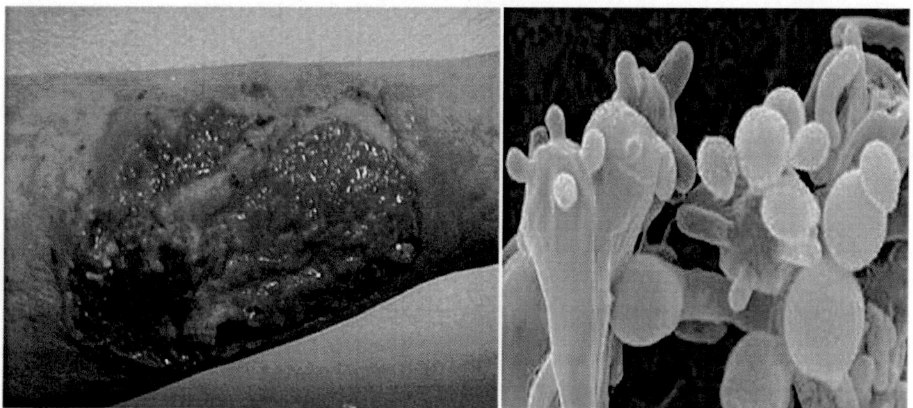

Fig. 172: Symptoms of *Cryptococcus gattii* infection on leg and the pathogenic yeast cells (in SEM).

The infection of Cryptococcal gattii can spreads from the lungs to the brain and this infection is known as Cryptococcal meningitis. The symptoms of cryptococcal meningitis include Headache, Fever, Neck pain, Nausea and vomiting, Sensitivity to light, Confusion or changes in behavior

4.7 Mycotic eye infections by fungi

Yeast Candida albicans, Filamentous fungi Fusarium solani and Aspergillus flavus.

Fungal eye infections (Fig.173) are extremely rare, but they can be very serious. The most common way for someone to develop a fungal eye infection is as a result of an eye injury, particularly if the injury was caused by fragments of dried plant material blown in the air. Inflammation or infection of the cornea (the clear, front layer of the eye) is known as **keratitis,** and inflammation or infection in the interior of the eye is called **endophthalmitis**.

The yeast *Candida albicans* is the most common cause of endogenous endophthalmitis. Filamentous fungi such as *Fusarium solani* and *Aspergillus flavus* may constitute up to one-third of cases of traumatic infectious keratitis. The symptoms of a fungal eye infection can appear anywhere from several days to several weeks after the fungi enter the eye.

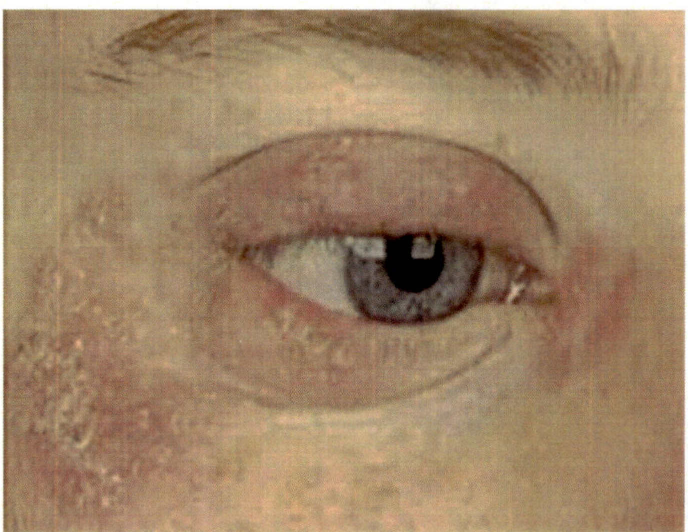

Fig. 173: Symptoms of fungal eye infection

The symptoms of a fungal eye infection are similar to those of other types of eye infections (such as those caused by bacteria) and can include Eye pain, Eye redness, Blurred vision, Sensitivity to light, Excessive tearing and Eye discharge.

4.8. Onychomycosis or Nail infections by fungi

Trichophyton rubrum

Fungal nail infections are common infections of the fingernails or toenails by the fungal pathogen *Trichophyton rubrum* that can cause the nail to become discolored, thick, and more likely to crack and break (Fig.174).

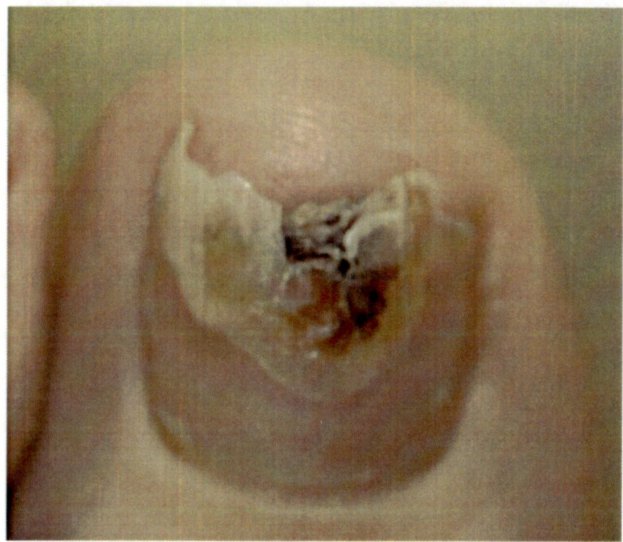

Fig. 174: Symptoms of fungal nail infection

Infections are more common in toenails than fingernails. A fungal nail infection usually is not painful unless it becomes severe.

4.9. *Histoplasmosis* disease causing fungi

Histoplasma

Histoplasmosis is an infection caused by a fungus called *Histoplasma* (Fig.175). The fungus lives in the environment, particularly in soil that contains large amounts of bird or bat droppings.

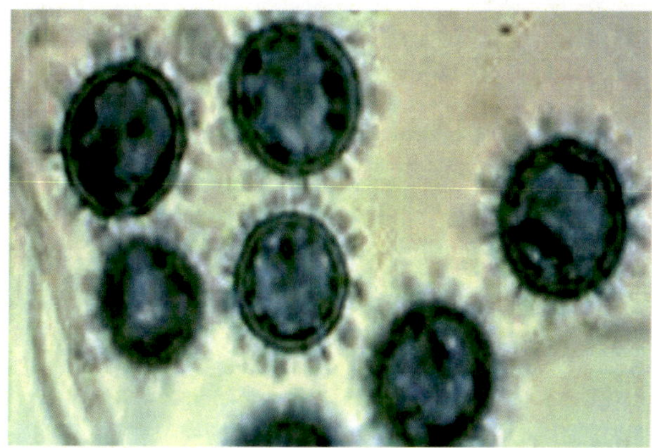

Fig. 175: Spores of *Histoplasma* fungi.causing Histoplasmosis.

Fever is a common symptom of histoplasmosis. Other people may have flu-like symptoms that usually go away on their own. Other Symptoms of histoplasmosis include Chills, Headache, Chest pain and Body aches.

People can get histoplasmosis after breathing in the microscopic fungal spores from the air. Although most people who breathe the spores do not get sick, those who do may have a fever, cough, and fatigue. Many people who get histoplasmosis will get better on their own without medication, but in some people, such as those who have weakened immune systems, the infection can become severe.

In the United States, *Histoplasma*is mainly present in the central and eastern states, especially areas around the Ohio and Mississippi River valleys. The fungus also lives in parts of Central and South America, Africa, Asia, and Australia.

4.10. *Mucormycosis* disease causing fungi

Mucorales

Mucormycosis also called zygomycosis is a rare infection caused by group of fungi called Mucoromycotina in the order Mucorales.These fungi are typically found in the soil and in association with decaying organic matter, such as leaves, compost piles, or rotten wood.

The symptoms of mucormycosis (Fig.176) depend on where in the body the fungus is growing. Mucormycosis most commonly affects the sinuses or lungs. Symptoms of sinus infections include fever, headache, and sinus pain. Lung infections with the fungus can cause fever and cough.

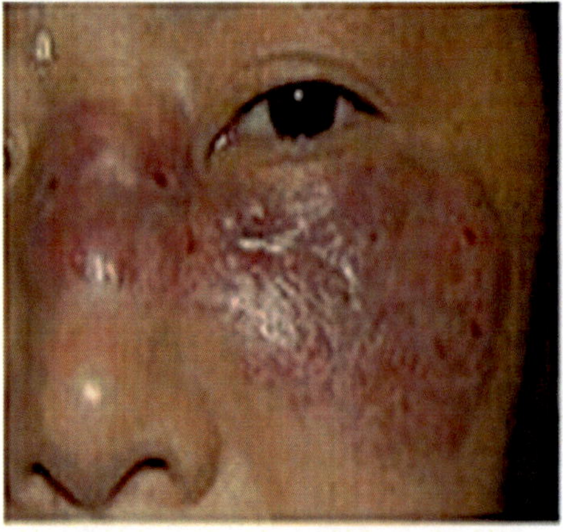

Fig. 176: Symptoms of *Mucormycosis* infection

Skin infections can develop after the fungus enters through a break in the skin caused by surgery, burns, or trauma. A skin infection can look like blisters or ulcers, and the infected tissue may turn black. Other symptoms of a skin infection include fever and tenderness, pain, heat, excessive redness, or swelling around a wound.

If the infection is not treated quickly, the fungus can spread throughout the body, and the infection is often fatal.

4.11. *Pneumocystis Pneumonia* disease causing fungi

Pneumocystis jirovecii

Pneumocystis pneumonia (PCP) is a serious illness caused by the fungus *Pneumocystis jirovecii* (Fig.177). PCP is one of the most frequent and severe opportunistic infections in people with weakened immune systems, particularly people with HIV/AIDS.

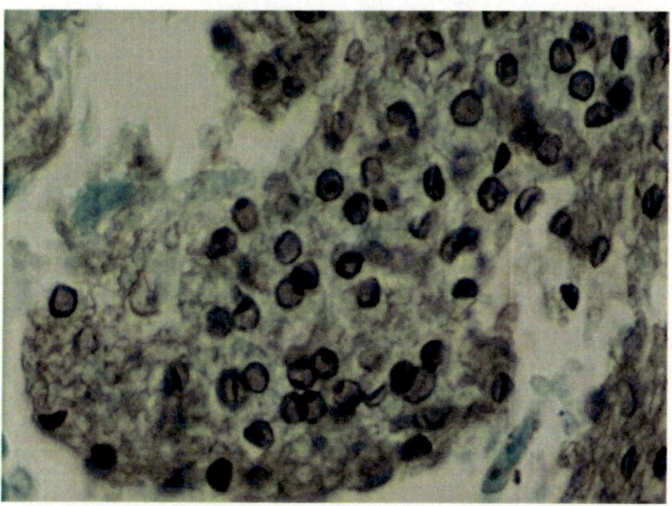

Fig. 177: Fungal spores of *Pneumocystis jirovecii* in PCP infection

The symptoms of PCP includes fever, dry cough, shortness of breath, and fatigue. In people with weakened immune systems, PCP can be very serious, so it is important to see a doctor if you have these symptoms. In HIV-infected patients, PCP usually presents sub-acutely, and symptoms include a low-grade fever.

4.12. *Ringworm* disease causing fungi

Trichopyton rubrum

Ringworm is a common infection of the skin and nails that is caused by fungus *Trichopyton rubrum* (Fig.178). The infection is called "ringworm" because it can cause an itchy, red, circular rash. Ringworm is also called "tinea" or "dermatophytosis". The different types of ringworm are usually named after the location of the infection on the body.

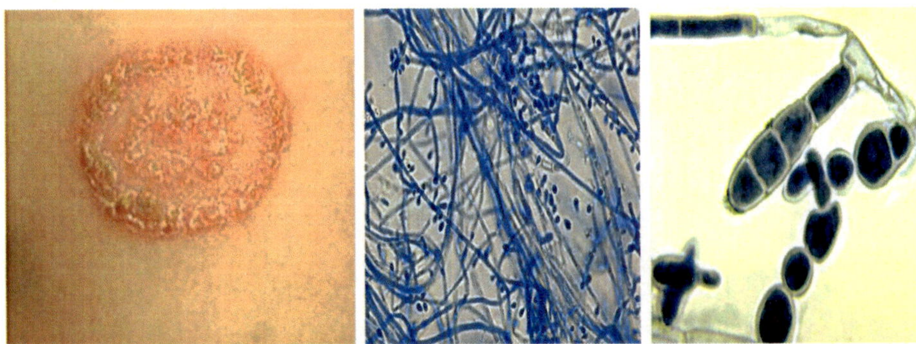

Fig. 178: Symptoms of ringworm infection and the structure of causal Fungus

Approximately 40 different species of fungi can cause ringworm; the scientific names for the types of fungi that causes ringworm are *Trichophyton, Microsporum, and Epidermophyto*

Ringworm can affect skin on almost any part of the body as well as fingernails and toenails. The symptoms of ringworm often depend on, which part of the body is infected, but the general symptoms include Itchy skin, Ring-shaped rash, Red, scaly, cracked skin and Hair loss.

Symptoms typically appear between 4 and 14 days after the skin comes in contact with the fungi that cause ringworm.

Symptoms of ringworm at different locations on the body

- **Feet (tinea pedis or "athlete's foot"):** The symptoms of ringworm on the feet include red, swollen, peeling, itchy skin between the toes (especially between the pinky toe and the one next to it). The sole and heel of the foot may also be affected. In severe cases, the skin on the feet can blister.
- **Scalp (tinea capitis):** Ringworm on the scalp usually looks like a scaly, itchy, red, circular bald spot. The bald spot can grow in size and multiple

spots might develop if the infection spreads. Ringworm on the scalp is more common in children than it is in adults.

- **Groin (tinea cruris or "jock itch"):** Ringworm on the groin looks like scaly, itchy, red spots, usually on the inner sides of the skin folds of the thigh.
- **Beard (tinea barbae):** Symptoms of ringworm on the beard include scaly, itchy, red spots on the cheeks, chin, and upper neck. The spots might become crusted over or filled with pus, and the affected hair might fall out.

4.13. *Sporotrichosis* disease causing fungi

Sporothrix schenckii

Sporotrichosis is an infection caused by a fungus called *Sporothrix schenckii (Fig.179)*. The fungus lives throughout the world in soil, plants, and decaying vegetation. Cutaneous (skin) infection is the most common form of infection and usually occurs after handling contaminated plant material, when the fungus enters the skin through a small cut or scrape.

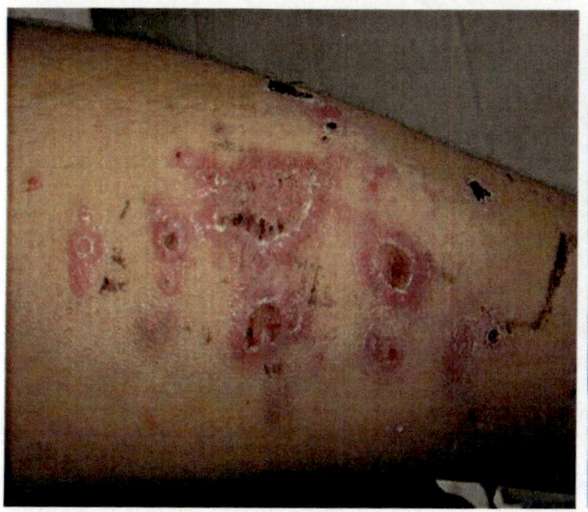

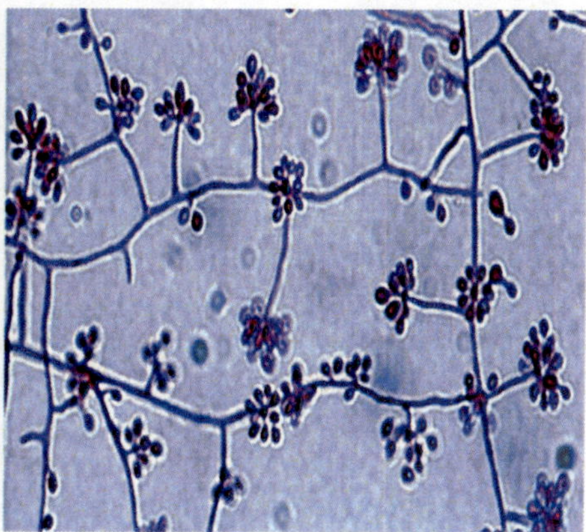

Fig. 179: Symptoms of sporotrichosis infection and the structure of causal Fungus

The first symptom is usually a small painless nodule (bump) resembling an insect bite. The first nodule may appear any time from 1 to 12 weeks after exposure to the fungus. The nodule can be red, pink, or purple in color, and it usually appears on the finger, hand, or arm where the fungus has entered through a break in the skin. The nodule will eventually become larger in size and may look like an open sore or ulcer that is very slow to heal. Additional bumps or nodules may appear later near the original lesion.

Most *Sporothrix* infections only involve the skin. However, the infection can spread to other parts of the body, including the bones, joints, and the central nervous system. Usually, these types of disseminated infections only occur in people with weakened immune systems. In rare cases, a pneumonia-like illness can occur after inhaling *Sporothrix* spores, which can cause symptoms such as shortness of breath, cough, and fever.

4.14. *Penicilliosis* disease causing fungi

Penicillium marneffei

Skin lesions are most commonly papules often with a central necrotic umbilication similar in appearance to those seen in molluscum contagiosum and are usually located on the face, trunk and extremities. In many cases numerous subcutaneous abscesses ulcerate over time (Fig.180).

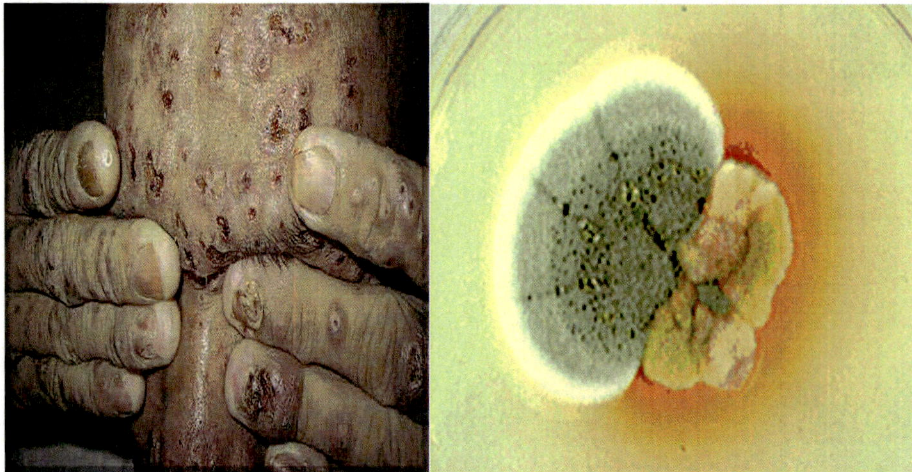

Fig. 180: Symptoms of *Penicilliosis* infection and the growth of causal Fungus

In patients with normal immunity, *P. marneffei* infection may either be disseminated or local. In the latter, both clinical and histological appearances may strongly resemble tuberculosis (eg. suppurative lymphadenopathy). In HIV patients, *P. marneffei* infection is usually disseminated at diagnosis. Commonly the skin, reticuloendothelial system, lung and gut are infected. Fungaemia is present in the majority of cases and other organ systems including kidney, bones, joints and pericardium may also be involved. Patients usually present with non-specific symptoms of fever, anaemia and weight loss. It is important to note that the clinical symptoms of disseminated infection by *P. marneffei* mimic those seen in AIDS patients with disseminated cryptococcal infection or disseminated histoplasmosis

To date, all naturally occurring infections have been in residents of, or travellers to, south east Asia; especially northern Thailand, Vietnam, Hong Kong, Taiwan and southern China. Imported cases of *P. marneffei* infections have been reported from Australia, France, Italy, Netherlands, UK and USA. Over 300 cases have been reported with the majority of these coming from Chiang Mai in northern Thailand. Other predisposing factors include lymphoproliferative disorders, bronchiectasis and tuberculosis, autoimmune diseases and corticosteroid therapy.

Penicillium marneffei exhibits thermal dimorphism by growing in living tissue or in culture at 37^{0}C as a yeast-like fungus or in culture at temperatures below 30^{0}C as a mould. It has a propensity to cause disease in the normal host, as well as in immune suppressed patients, but significantly, it has now become a major opportunistic pathogen in HIV positive patients in Indochina.

4.15. *Phaeohyphomycosis* disease causing fungi

Phaeohyphomyces

Clinical forms of phaeohyphomycosis range from localized superficial infections of the stratum corneum (tinea nigra) to subcutaneous cysts (phaeomycotic cyst) to invasion of the brain. Ideally, individual disease states involving an invasive fungal infection by a dematiaceous hyphomycete should be designated by a specific description of the pathology and the causative fungal genus or species (where known).

The etiological agents include various dematiaceous hyphomycetes especially species of *Exophiala, Phialophora, Wangiella, Bipolaris, Exserohilum, Cladophialophora, Phaeoannellomyces, Aureobasidium, Cladosporium, Curvularia* and *Alternaria.* Ajello (1986) listed 71 species from 39 genera as causative agents of phaeohyphomycosis

Clinical manifestations

1. Subcutaneous phaeohyphomycosis

Subcutaneous infections occur worldwide, usually following the traumatic implantation of fungal elements from contaminated soil, thorns or wood splinters. *Exophiala jeanselmei* and *Wangiella dermatitidisare* the most common agents and cystic lesions (Fig.181) occur most often in adults. Occasionally, overlying verrucous lesions are formed, especially in the immune suppressed patient.

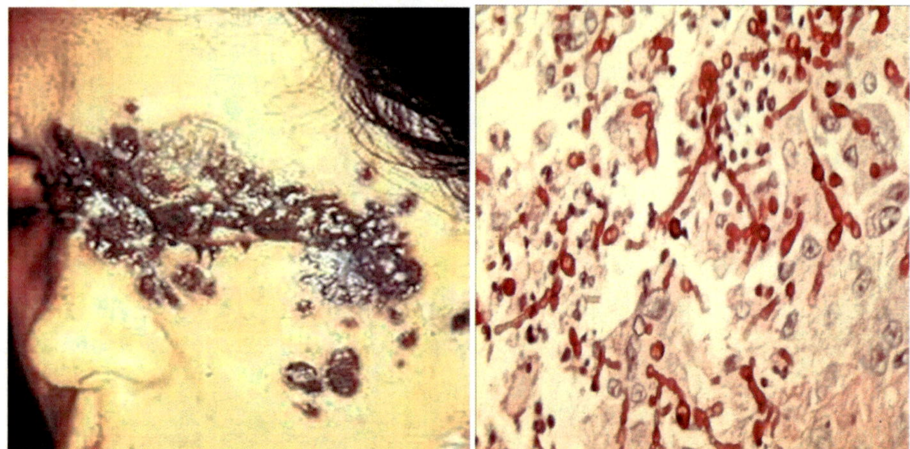

Fig. 181: Symptoms of *Phaeohyphomycosis* infection and the structure of causal Fungus

2. Paranasal sinus phaeohyphomycosis

Sinusitis caused by dematiaceous fungi, especially species of *Bipolaris, Exserohilum, Curvularia* and *Alternaria* is increasingly being reported, especially in patients with a history of allergic rhinitis or immunosuppression.

3. Cerebral phaeohyphomycosis

Cerebral phaeohyphomycosis is a rare infection, occurring mostly in immune suppressed patients following the inhalation of conidia. However, cerebral infections caused by *Cladophialophora bantiana* have been reported in a number of patients without any obvious predisposing factors. This fungus is neurotropic and dissemination to sites other than the CNS is rare.

4.16. *Hyalohyphomycosis* disease causing fungi

Acremonium, Paecilomyces etc

A mycotic infection of man or animals caused by a number of hyaline (non-dematiaceous) hyphomycetes where the tissue morphology of the causative organism is mycelial. This separates it from phaeohyphomycosis where the causative agents are brown-pigmented fungi. Hyalohyphomycosis is a general term used to group together infections caused by unusual hyaline fungal pathogens that are not agents of otherwise-named infections; such as Aspergillosis.

Etiological agents include species of *Penicillium, Paecilomyces, Acremonium, Beauveria, Fusarium* and *Scopulario*

The clinical manifestations of hyalohyphomycosis are many ranging from harmless saprophytic colonization to acute invasive disease (Fig.182).

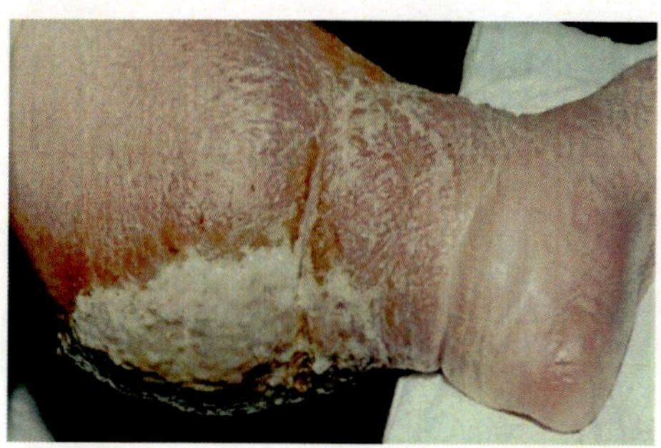
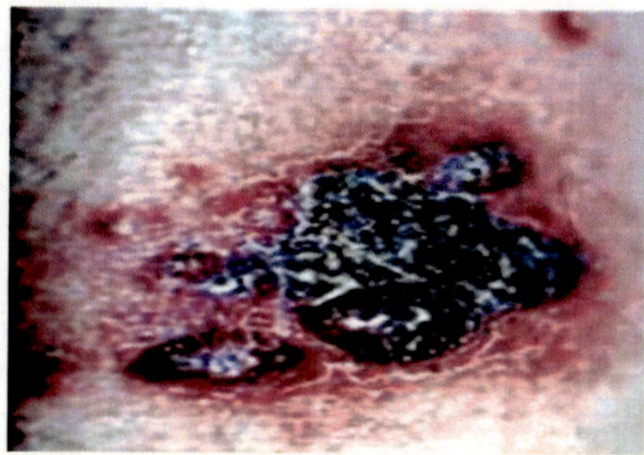

Fig. 182: Symptoms of *Hyalohyphomycosis* infection

Ideally, individual disease states involving invasive fungal infection by a hyaline hyphomycete should be designated by specific description of the pathology and the causative fungal genus or species (where known). Predisposing factors include prolonged neutropenia, especially in leukemia patients or in bone marrow transplant recipients, corticosteroid therapy, cytotoxic chemotherapy and to a lesser extent patients with AIDS. The typical patient is granulocytopenic and receiving broad-spectrum antibiotics for unexplained fever.

5

Fungi Causing Diseases in Animal

5.1. *Ringworm* (in Cattle) causing fungi

Trichophyton verrucosum

Ringworm is one of the commonest skin diseases in cattle. Ringworm is a transmissible skin infectious disease caused most often by *Trichophyton verrucosum*, a spore forming fungi. The spores can remain alive for years in a dry environment. Although unsightly, fungal infections cause little permanent damage or economic loss. Direct contact with infected animals is the most common method of spreading the infection.

The symptoms of Ringworm include Lesions with greywhite areas with an ash like surface usually circular in outline and slightly raised that appears on skin. Size of lesions vary and can become very extensive. In calves the infection is most commonly found around eyes, on ears and on back. In adult cattle infection on chest and legs are more common (Fig.183)

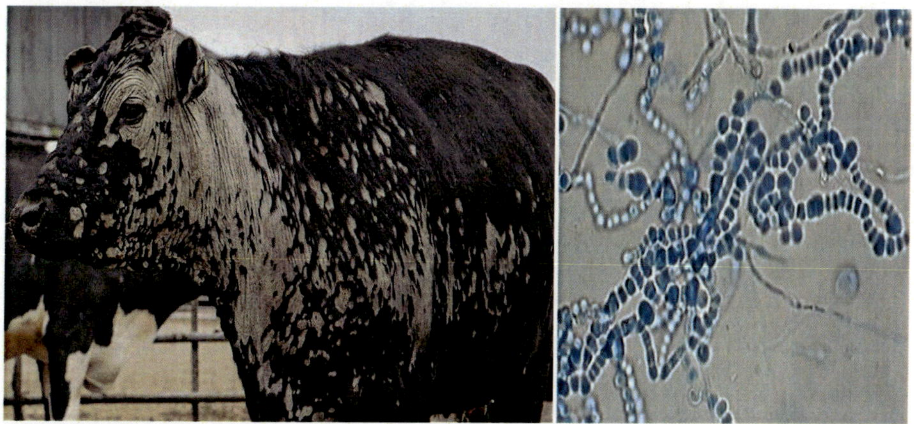

Fig. 183: Symptoms of ringworm infection and the pathogen *Trichophyton*

5.2. *Facial Eczema* (in Sheep and Cattle) causing fungi

Pithomyces chartarum

Facial eczema is a type of sunburn (photosensitisation) affecting exposed areas of pale skin of sheep and cattle. It is caused by a poisonous substance called "sporidesmin", which is produced on pasture plants by the fungus ***Pithomyces chartarum***. In Victoria, the disease has occurred mainly in East Gippsland between Rosedale and Bairnsdale. Major outbreaks occurred in sheep in 1956 and 1959, and in cattle on irrigated pasture in 1974. Sporadic cases, mainly in sheep, have been recorded in other parts of Victoria.

The disease in stock may be seen between 7 and 20 days of pick-up of the toxic spores from the pasture. The toxin is absorbed from the intestine and reaches the liver, where it causes severe damage, first to the bile ducts and then to the liver cells themselves. All the outward signs of facial eczema result from the liver damage caused by sporidesmin

The disease signs in Sheep range from mild photo sensitisation to severe jaundice and death, depending on the amount of sporidesmin consumed. Sunburn is the most consistent sign, and usually affects the exposed areas of the skin of the face, ears, teats, and vulva. The skin over these areas becomes reddened, and then goes crusty and dark (Fig.184). It eventually peels off leaving large raw areas, which are susceptible to infections and flystrike. The sunburn is often accompanied by watery swelling of the underlying tissues, resulting in drooping ears and puffy eyes and face. Jaundice (yellowing of mucous membranes) is often seen at this stage. The animals lose weight rapidly. Most animals recover from the acute phase, but they may take many months to regain condition and wool growth. Some never recover, and either die or are culled.

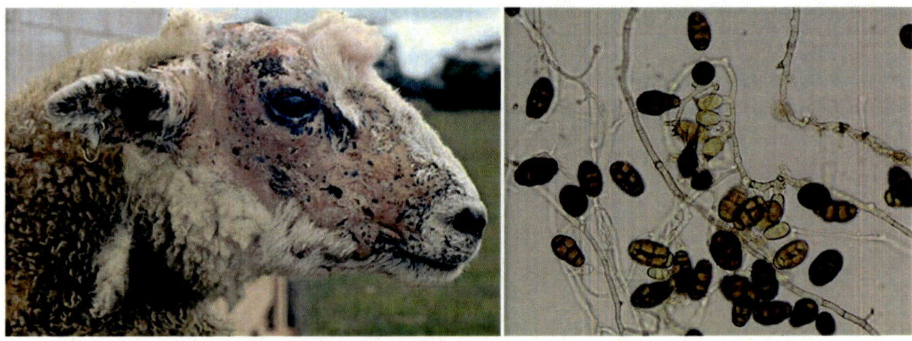

Fig. 184: Symptoms of *Facial eczema* infection and the pathogen pithomyces

The disease signs in Cattle are similar to those in sheep, but the sunburn also affects areas of skin lacking dark pigmentation (that is, areas covered by white hair) as well as exposed skin. In dairy cattle, the udder and teats are often severely affected, and milk production drops sharply. Loss of weight and general illness are often severe in cattle, and death, although uncommon, can occur up to months after the initial liver damage occurs.

5.3. *Aspergillosis* (in Dogs) causing fungi

Aspergillus sp

Aspergillosis is an opportunistic fungal infection caused by the fungus *Aspergillus*, a species of common mould found throughout the environment, including dust, straw, grass clippings, and hay. An "opportunistic infection" occurs when an organism, which does not generally cause disease, infects a dog. However, in the case of aspergillosis, it does because the pet's immune system and/or body is weakened from some other disease.

There are two types of *Aspergillus* infection i.e., nasal and disseminated. Both types can occur in cats and dogs, but they occur more frequently in dogs. Young adult dogs with a long head and nose (known as dolichocephalic breeds) and dogs with a medium length head and nose (known as mesatcephalic breeds) are also more susceptible to the nasal form of aspergillosis. The disseminated version of the disease seems to be more common in German Shepherds.

The nasal type of infection is localized in the nose, nasal passages, and front sinuses (Fig.185). It is believed that this develops from direct contact with the fungus through the nose and sinuses. For example, if an animal is outside and around dust and grass clippings, the fungus may enter via the moist lining of the nose.

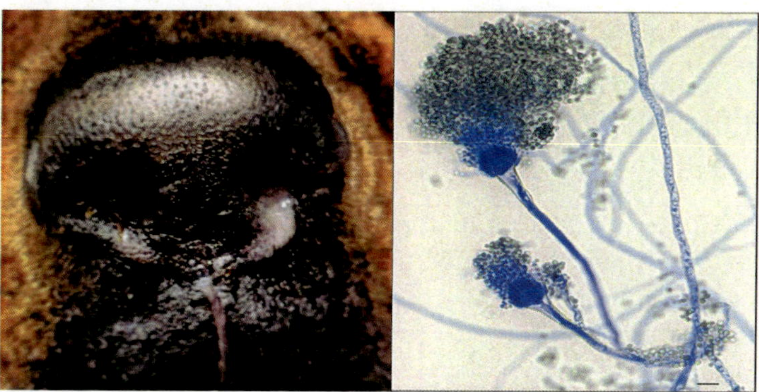

Fig.185: Symptoms of Aspergillosis in dog and the structure of *Aspergillus* pathogen

Symptoms of nasal aspergillosis include sneezing, nasal pain, bleeding from the nose, reduced appetite, visibly swollen nose, and long-term nasal discharge from the nostril(s), which may contain mucus, pus and/or blood. In some cases, loss of pigment or tissue on the surface of the skin may also occur.

The second type of *Aspergillus* infection is disseminated, meaning it is more widespread, and is not only located in the nasal area. It's not certain how this form enters the body. Symptoms of disseminated aspergillosis in dogs may develop suddenly or slowly over a period of several months, and include spinal pain or lameness due to infection, and cause inflammation of the animal's bone marrow and bones. Other signs which are not specific to the disease include fever, weight loss, vomiting, and anorexia.

5.4. *Blastomycosis* (in Dogs) causing fungi

Blastomyces dermatitidis

Blastomycosis is a systematic yeast like fungal infection caused by the organism *Blastomyces dermatitidis*, which is commonly found in decaying wood and soil. *Blastomycosis* (Fig.186) occurs most frequently in male dogs, but female dogs are also susceptible.

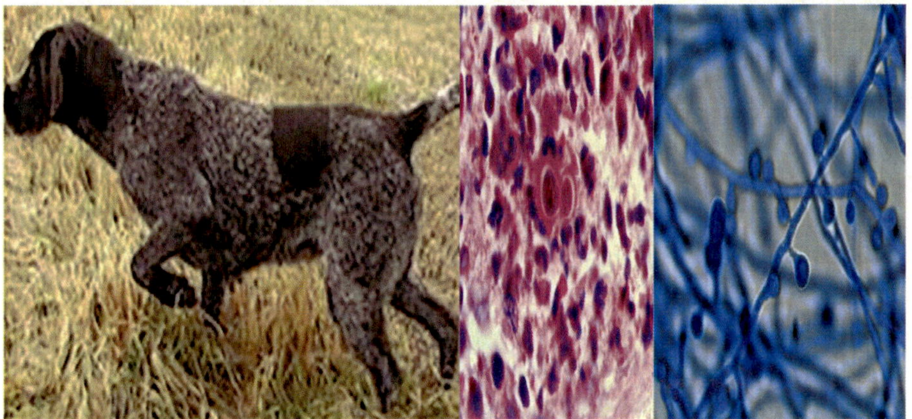

Fig.186: Symptoms of *Blastomycosis* in dog and the structure of Blastomyces pathogen

The symptoms includes Fever, Loss of appetite (anorexia), Weight loss, Eye discharge, Eye inflammation, specifically the iris, Difficulty in breathing (e.g., coughing, wheezing and other unusual breathing sounds) and Skin lesions, which are frequently filled with pus

6

Fungi Causing Diseases in Birds

6.1. *Aspergillosis* disease causing fungi

Aspergillus fumigatus

Aspergillosis is a fungal infection that commonly causes respiratory disease in pet birds. It can infect both upper (nose, sinuses, eye, and trachea) and lower (lung and air sacs- a specialized part of the respiratory tract that birds have) respiratory organs or more broadly distributed systemic infections (Fig.187).

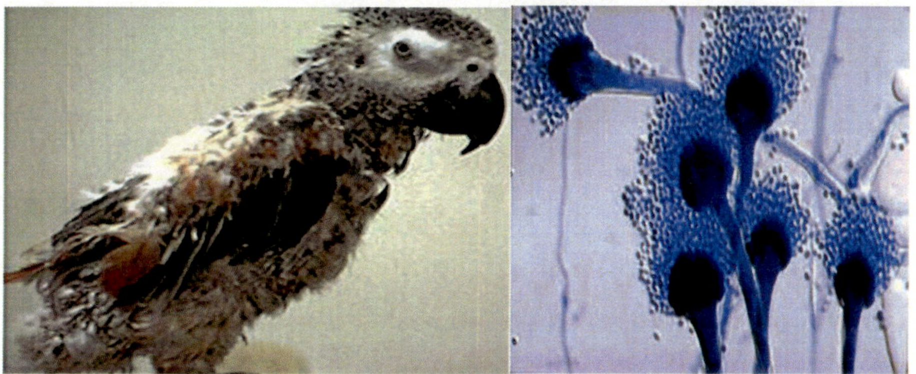

Fig.187: Symptoms of *Aspergillosis* in bird and the structure of Aspergillus pathogen

The Symptoms includes Weight loss, Open-mouth breathing or gasping for breath, Tracheal movement or in-out-in movements in the region of the neck, Tail-bobbing or movement of the bird's tail up and down while breathing in and out, heavy breathing after flight or exercise, Change in voice and persistent itching or rubbing of cerenares.

6.2. Candidiasis disease causing fungi

Candida albicans

Clinical signs of candidiasis vary depending upon the location of infection. Local infections within the oropharynx may cause difficulty or reluctance to swallow food and halitosis. Oropharyngeal infections commonly result in psuedomembranes/plaques of necrotic debris overlying the mucous membranes or catarrhal inflammation giving the mucous membranes a "Turkish towel" appearance. Infection within the crop may result in regurgitation, vomiting, delayed crop emptying, anorexia, palpable thickening of the crop and ingluviolith formation (Fig.188). Proventricular and ventricular infections may cause vomiting, weight loss, diarrhea and general malaise.

Fig.188: Symptoms of Candidiasis in bird and the structure of *Candida albicans* (in SEM) pathogen

Candida sp may also colonize the respiratory tract leading to dyspnea, ocular or nasal discharge and sneezing. Less commonly, *Candida* sp may also infect ocular tissues, skin, bone marrow, liver, pericardial tissues and the CNS (canaries). *Candida parasilosis* has been reported to cause systemic infection of the bone marrow and liver as well as splenic degeneration.

6.3. *Cryptococcosis* disease causing fungi

Cryptococcus neoformans

Clinical signs of *Cryptococcus* infections may be non-specific and include weakness, lethargy, depression, anorexia, diarrhea, dyspnea, weight loss, blindness and paralysis (Fig.189).

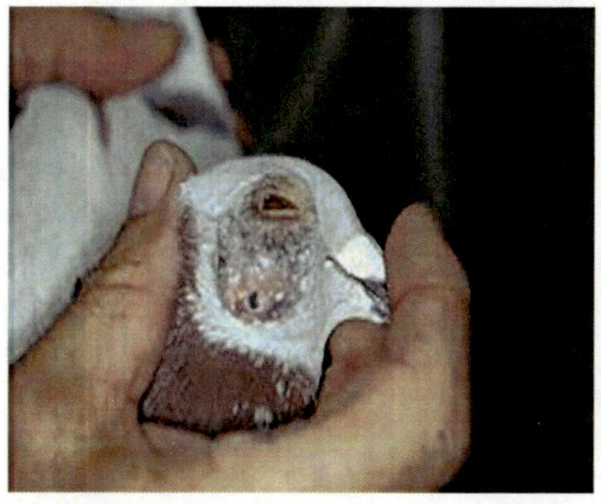
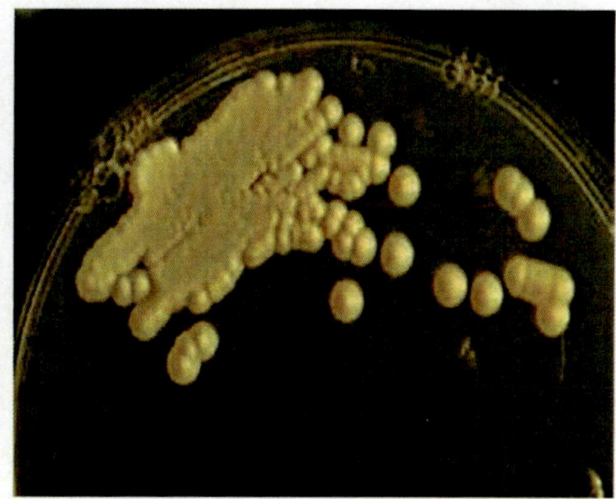

Fig.189: Symptoms of *Cryptococcosis* in bird and the colonies of fungal pathogen

Neurologic signs may occur with CNS (brain and meningeal) involvement. Moderate to severe dyspnea may be seen with involvement of both the upper or lower respiratory tract.

6.4. *Rhodotoruliasis* disease causing fungi

Rhodotorula mucilaginosa

Rhodotorula mucilaginosa is a yeast that infects the skin and is occasionally seen in raptors (especially falcons). The pathogen typically causes greasy, yellowish-brown crust overlying cracked and discoloured areas of skin in the axillary area of the wings (Fig.190) or between the thigh musculature and the body wall. Without treatment, lesions may become hyperkeratotic resulting in proliferative, horny growths on the effected skin.

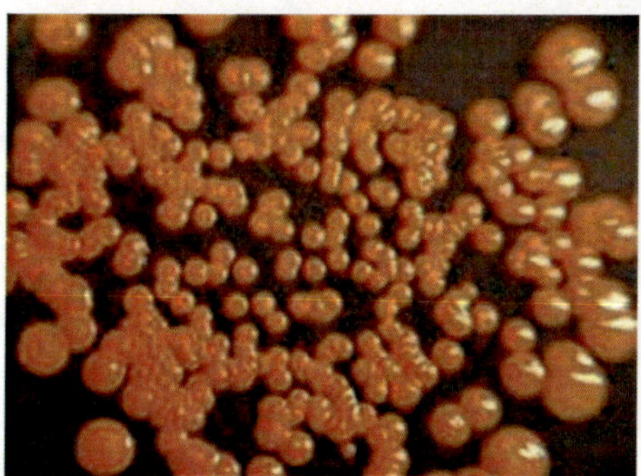

Fig. 190: Symptoms of *Rhodotoruliasis* in bird and the colonies of fungal pathogen

6.5. Mucormycosis disease causing fungi

Mucor spp, *Absidia* spp *and Rhizopus* spp

The saprophytic fungi of these group causes disease of varying forms depending upon the organ system effected. Clinical signs of disease may be the result of enteritis, air sacculitis, osteomyelitis, myocarditis and nephritis. Mycotic infections due to *Mucor* sp have been reported in African grey parrots.

6.6. White Nose Syndrome causing fungi

Pseudogymnoascus destructans

Since 2006, White-nose Syndrome (WNS) has devastated bat populations across Eastern North America, causing the most precipitous wildlife decline of the past century. WNS has killed at least 5.7 million bats since it was discovered in a single cave in New York. Today, seven bat species in 26 states and 5 Canadian provinces have been diagnosed with the devastating disease. The disease is named for a cold-loving white fungus *Pseudogymnoascus destructans* typically found on the faces and wings of infected bats (Fig.191).

Fig.191: Symptoms of WNS in bird and the structure of fungal pathogen (In SEM)

White-nose Syndrome causes bats to awaken more often during hibernation and use up the stored fat reserves that are needed to get them through the winter. Infected bats often emerge early from hibernation and are often seen flying around in midwinter. These bats usually freeze or starve to death.

The Symptoms observed with WNS include unusual winter behavior like abnormally frequent or abnormally long arousal from torpor (temporary hibernation), flying, loss of body fat, damage and scarring of the wing membranes, and death.

7

Fungi Causing Diseases in Aquatic Animal/Fish

7.1. Black Gill Disease (in Shrimps) causing fungi

Fusarium spp

The affected prawns show brown to black spots on the gills. In acute cases gills may completely become brown or black with atrophy and necrosis (Fig.192).

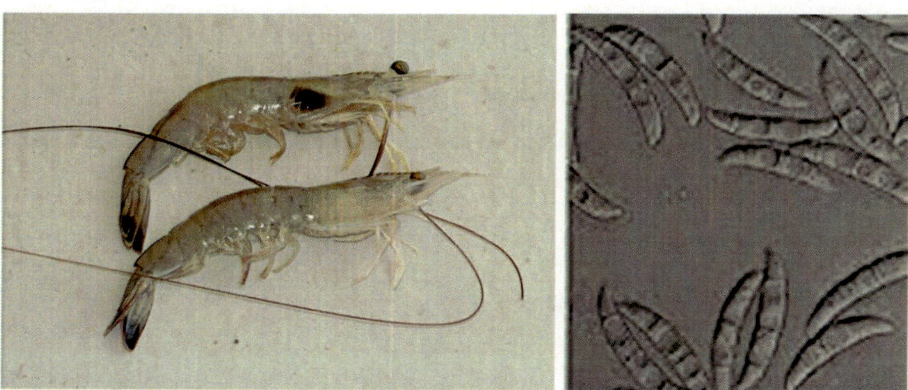

Fig. 192: Symptoms of black gill disease and the structure of *Fusarium* spores.

7.2. *Saprolegniasis* disease causing fungi

Saprolegnia

Saprolegniasis is a fungal disease of fish and fish eggs, most commonly caused by the *Saprolegnia* species called "water molds." They are common in fresh or brackish water. *Saprolegnia* can grow at temperatures ranging from 0° to 35°C but seem to prefer temperatures of 15° to 30°C. The disease fungus attack an existing injury on the fish and spread to healthy tissue. Poor water quality (for example, water with low circulation, low dissolved oxygen, or high ammonia) and high organic loads, including the presence of dead eggs, are often associated with *Saprolegnia* infections.

Saprolegniasis (Fig.193) is often first noticed by observing fluffy tufts of cottony white to grey and brown fungal growth on skin, fins, gills, or eyes of fish or on fish eggs. With progression of infection fish usually becomes lethargic and less responsive to external stimuli. So fish under such conditions is a target to predators

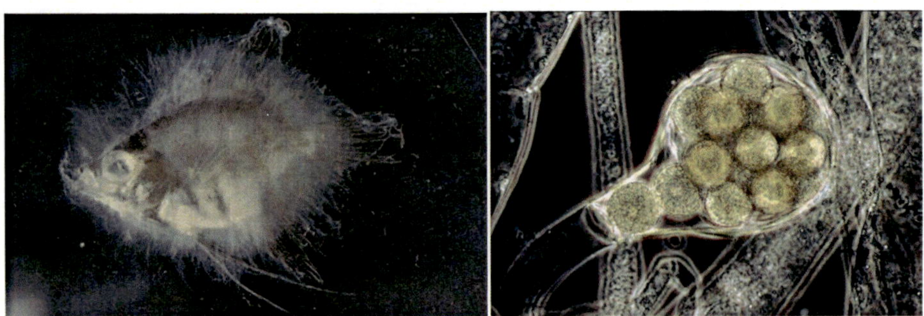

Fig. 193: Symptoms of *Saprolegniasis* and the structure of Saprolegnia fungus and spore sac.

7.3. *Branchiomycosis* disease causing fungi

Branchiomyces demigrans and Branchiomyces sanguinis

Branchiomycosis or "Gill Rot" (Fig.194) is caused by the fungi *Branchiomyces sanguinis* (carps) and *Branchiomyces demigrans* (Pike and Tench).

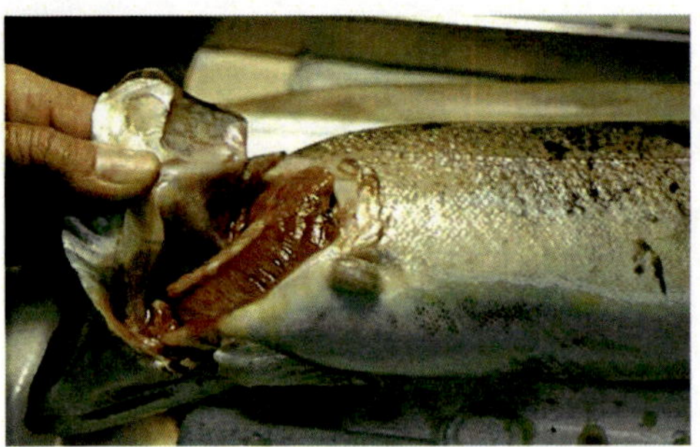

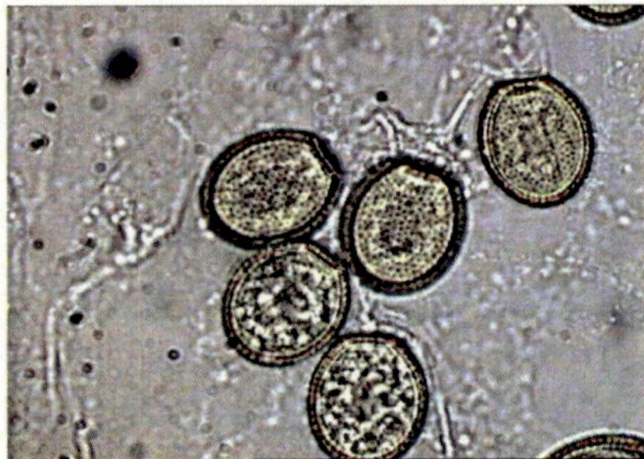

Fig. 194: Symptoms of *Branchiomycosis* and the pathogenic fungus spore.

Branchiomycosis is a pervasive problem in Europe, but has been only occasionally reported by U.S. fish farms. Both species of fungi are found in fish suffering from an environmental stress, such as low pH (5.8 to 6.5), low dissolved oxygen, or a high algal bloom. *Branchiomyces* sp. grow at temperatures between 13.8° and 35°C but grow best between 25° and 32.2°C.

The main sources of infection are the fungal spores carried in the water and detritus on pond bottoms.

Branchiomyces sanguinis and *B. demigrans* infect the gill tissue of fish. Fish may appear lethargic and may be seen gulping air at the water surface (or piping). Gills appear striated or marbled with the pale areas representing infected and dying tissue. As the tissue dies and falls off, the spores are released into the water and transmitted to other fish. High mortalities are often associated with this infection.

7.4. Swinging disease / *Icthyophonus fungal* infection

Icthyophonus hoferi

The Icthyophonus disease fungus, *Icthyophonus hoferi* grows in fresh and saltwater, in wild and cultured fish, but is restricted to cool temperatures (2.2° to 20°C). The disease is spread by fungal cysts which are released in the faeces and by cannibalism of infected fish.

Because the primary route of transmission is through the ingestion of infective spores, fish with a mild to moderate infection will show no external signs of the disease. In severe cases, the skin may have a "sandpaper texture" caused by infection under the skin and in muscle tissue (Fig.195). Some fish may show curvature of the spine. Internally, the organs may be swollen with white to grey-white sores.

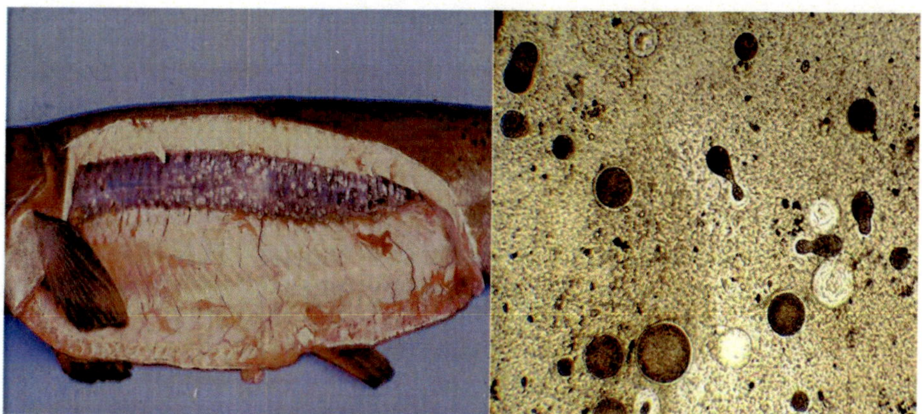

Fig. 195: Symptoms of Swinging disease and the pathogenic fungus spore cysts

Diseased fish shows curious swinging movements hence the disease is called as swinging disease. Along with liver severely affected organs are spleen (in salmonids), heart (in herring), kidney (in salmonids), gonads,brain (in salmonids), gills (in salmonids), and musculature and nerve tissue behind the eyes.

7.5. *Aspergillosis in corel*

Aspergillus sydowii

This disease appears as white or purple spots randomly appearing all over the coral (Fig.196). As these spots get bigger, the disease kill the tissue of the corals. It mostly affects soft corals originating from the Caribbean Sea and is very common in sea fans and sea whips.

Fig. 196: Symptoms of *Aspergillosis* in Corel and the structure of *Aspergillus* pathogen

7.6. *Chytridiomycosis* disease in Amphibian causing fungi

Batrachochytrium dendrobatidis

Chytridiomycosis is an emerging infectious disease of amphibians caused by an aquatic fungal pathogen, *Batrachochytrium dendrobatidis* (*Bd*) (Fig.197). Chytridiomycosis disease occurs when an amphibian is infected with large numbers of the *Bd* fungus. Infection with *Bd* occurs inside the cells of the outer skin layers that contain large amounts of a protein called "keratin". Keratin is the material that makes the outside of the skin tough and resistant to injury and is also what hair, feathers and claws are made of. With chytridiomycosis, the skin becomes very thick due to a microscopic change in the skin that pathologists call "hyperplasia and hyperkeratosis". These changes in the skin are deadly to amphibians because unlike most other animals, amphibians "drink" water and absorb important salts (electrolytes) like sodium and potassium through the skin and not through the mouth. Abnormal electrolyte levels as the result of *Bd* damaged skin, cause the heart to stop beating and the death of the animal

(Voyles et al., 2009). Other amphibians like the lungless salamanders, use the skin to breathe and skin changes due to chytridiomycosis could interfere with this function causing suffocation.

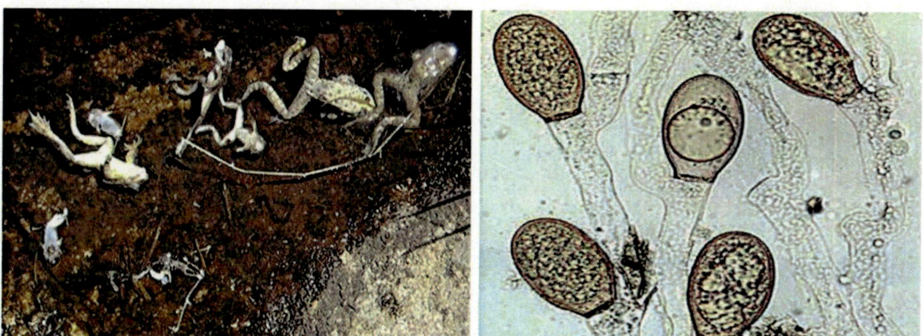

Fig.197: Symptoms of *Chytridiomycosis* in Amphibian and the structure of *Batrachochytrium dendrobatidis* pathogen

An amphibian that is sick with chytridiomycosis have a wide variety of symptoms or "clinical signs". Some of the most common signs are reddened or otherwise discolored skin, excessive shedding of skin, abnormal postures such as a preference for keeping the skin of the belly away from the ground, unnatural behaviors such as a nocturnal species that suddenly becomes active during the day, or seizures. Many of these signs are said to be "non-specific" and many different amphibian diseases have signs that overlap with those of chytridiomycosis. In addition, other cases of chytridiomycosis will not show any of these signs and amphibians will simply be found dead.

Section. III
Useful Fungi

8

Fungi Useful in Industrial Production

8.1. Fungi useful in food processing

8.1.1. Aspergillus sojae and Aspergillus oryzae

Aspergillus sojae and *Aspergillus oryzae* (Fig.198) is a mold species in the genus *Aspergillus*. Its use was originated in China in the 2nd century AD and spread throughout East Asia where it is used in cooking and as a condiment.

Uses

- It is used to prepare Soy sauce (also called **soya sauce**) a condiment made from a fermented paste of boiled soybeans, roasted grain and brine,
- In Japan it is used to make the ferment (Kōji) of soy sauce, the mirin and other lacto-fermented condiments like tsukemono. Soy sauce is a condiment produced by fermenting soybeans with *Aspergillus sojae*, along with water and salt.

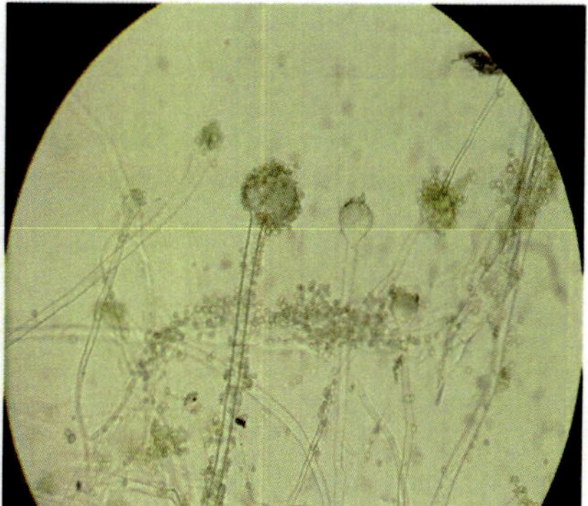

Fig. 198: *Aspergillus* sp. used in food processing.

8.1.2. Mucor species

Mucor spp, *Actinomucor elegance* and other 6 species of this moulds are used in the preparation of fermented tofu (Fig.199) also known as fermented bean curd, lufu, rufu, sufu, tofu cheese, soy cheese etc . Preserved tofu is a form of processed, preserved tofu used in East Asian cuisine as a condiment made from soybeans.The ingredients typically are soybeans, salt, rice wine and sesame oil or vinegar with the Mucorales fungal species.

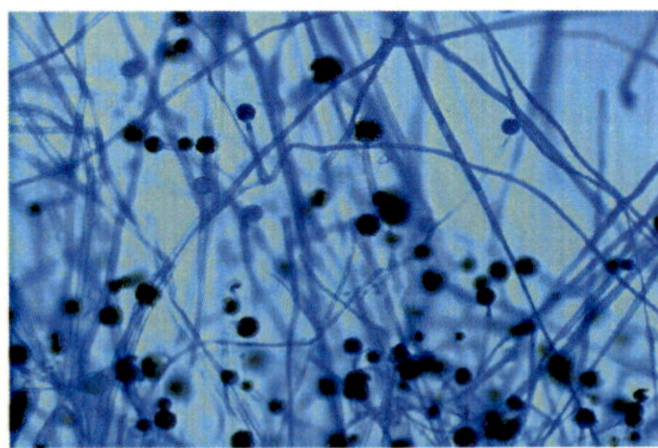

Fig.199. Fermented Tofu and structure of Mucor sp used in its processing.

Use

- Fermented tofu is commonly used as a condiment, combined in to sauces to accompany hot pot, or consumed at breakfast to flavor rice, porridge, gruel, congee, or erkuai.

8.1.3. Rhizopus oligosporus

Rhizopus oligosporus is used in the preparation of homemade Indonesian food Tempeh (Fig.200).

Fig.200: Processed Tempeh and structure of *Rhizopus oligosporus* used in its processing.

Use

- It is used in preparation of popular Indonesian food, Tempeh, which is created by fermenting soybeans in combination with *Rhizopus oligosporus.*

8.2. Fungi useful in Brewing and Bakery production

8.2.1. Sacchromyces cerevisiae

The fungal yeast *Sacchromyces cerevisiae* (Fig.201) is used in various bakery and brewerage production

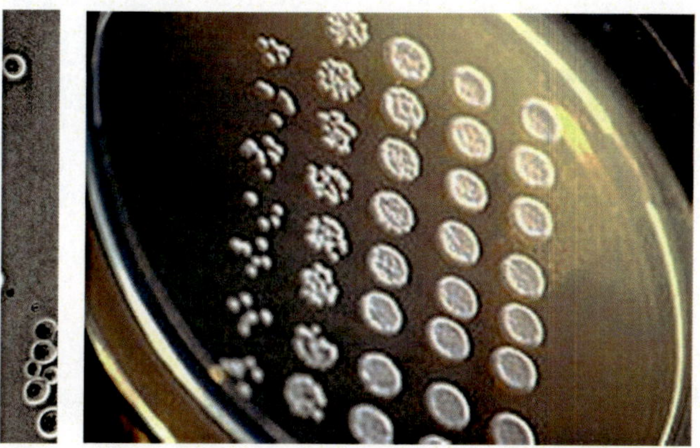

Fig. 201: Fungal yeast colonies of *Sacchromyces cerevisiae* on an agar plate and structure of yeast cells.

Uses In

• Candies Preparation

Saccharomyces cerevisiae is used in preparation of soft-center candies., eg cordial cherries.

• Brewing

Saccharomyces cerevisiae is used in brewing beer, when it is sometimes called a top- fermenting or top-cropping yeast. Also various species of *saccharomyces* are used in natural fermentation of wines ake, ale, brandy, whiskey, rum, vodka and tequila.

• Baking

S. cerevisiae is used in baking. The carbon dioxide generated during the fermentation act as a leavening agent in bread and other baked goods. Historically, its use was closely linked to the brewing industry's use of yeast, as bakers took or bought the barm or yeast- filled foam from brewing ale from the brewers (producing the barm cake); today, brewing and baking yeast strains are somewhat different.

• Uses in aquaria

Owing to the high cost of commercial CO_2 cylinder systems, CO_2 injection by yeast is one of the most popular DIY approaches followed by aquaculturists for providing CO_2 to underwater aquatic plants. The yeast culture is, in general, maintained in plastic bottles, and typical systems provide one bubble every 3–7 seconds. Various approaches have been devised to allow proper absorption of the gas into the water.

8.3. Fungi used in cheese production

8.3.1. *Penicillium roquefortii*

Penicillium roqueforti (Fig.202) is a common saprotrophic fungus widespread in nature, can be isolated from soil, decaying organic matter, and plants. This fungus is used in the cheese production.

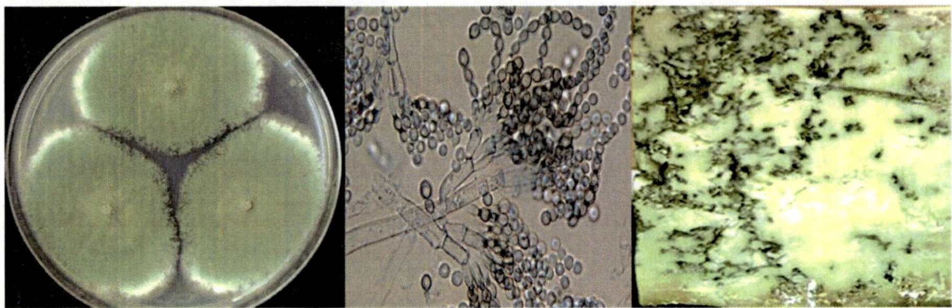

Fig. 202: Fungal culture, structure of *Penicillin roquefortii* and processed blue cheese

Use

1. Used in production of Blue cheese.

8.3.2. Penicillium camemberti

Penicillium camemberti (Fig.203) is used in the production of Camembert, Brie, Coulommiers and Cambozola cheeses, on which colonies of

P. camemberti form a hard, white crust. It is responsible for giving these cheeses their distinctive taste.

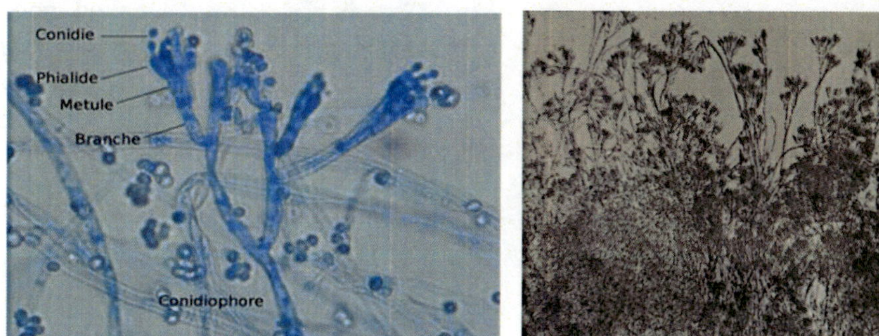

Fig. 203: Fungal structure *Penicillium camemberti* used in Cambozola cheeses production

Use

- *P. camemberti* is responsible for the soft, buttery texture of Brie and Camembert, but a too high concentration may lead to an undesirable bitter taste.

8.3.3. *Penicillium glaucum*

Penicillin glaucum is used in preparation on Gorgonzola Cheese (Fig.204).

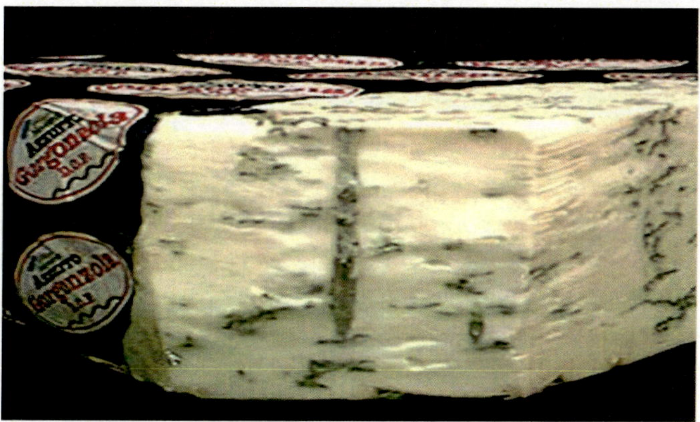

Fig. 204: Gorgonzola, an Italian cheese containing sheds of *Penicillium glacum*

8.3.4. Geotrichum candidum

Geotrichum candidum (Fig.205) is used widely in the production of certain dairy products including rind cheeses such as Camembert, Saint-Nectaire, Reblochon and others.

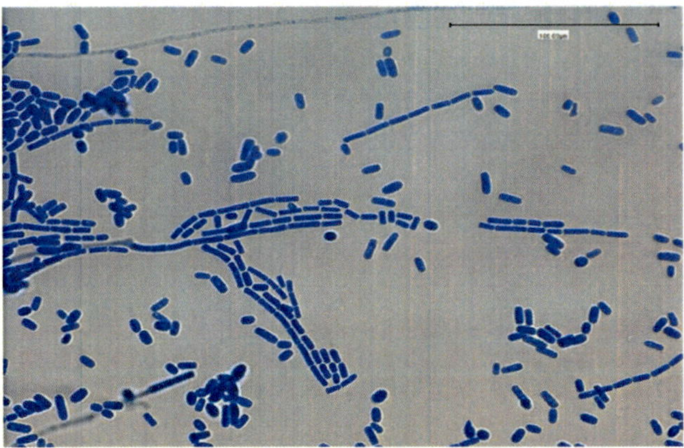

Fig. 205: Fungal structure of *Geotrichum candidum* in cheese and other products

Uses

- The fungus can also be found in a Scandinavian yogurt-like product known as viili where it is responsible for the product's velvety texture.
- *Geotrichum candidum* can be used commercially to inoculate wash-rinds and bloomy rind cheeses. Cultures can be added to milk, brine or sprayed onto cheese surface. The optimum pH range for growth on cheese ranges from 4.4 to 6.7. The fungus colonizes nearly the entire surface of the cheese during the early stages of ripening. It is found on soft cheeses like Camembert cheese and semi-hard cheese Saint-Nectaire and Reblochon. For the Camembert cheese the fungi grows on the outside of the cheese forming a rind.
- The fungus is responsible for the uniform, white, velvety coat on Saint Marcellin cheese. Lipases and proteases from *G. candidum* release fatty acids and peptides that provide the cheese with distinctive flavors. *Geotrichum candidum* reduces the bitterness in Camembert cheese through the activity of the aminopeptidases that hydrolyze low molecular weight hydrophobic peptides. Aminopeptidases also contributes an aroma in traditional Norman Camembert. The fungus also neutralize the curd by catabolizing lactic acid produced by bacteria. *Geotrichum candidum* prepares the cheese for colonization of other acid sensitive bacteria such as *Brevibacterium*. The fungus inhibits growth of the bacteria *Listeria monocytogenes*. Commercial strains of *G. candidum* are available for cheese ripening.

8.4. Fungi used in contamination detecting kits for food items

Penicillium digitatum

Penicillium digitatum is used as a biological tool during the commercial production of latex agglutination kits. Latex agglutination detects *Aspergillus* and *Penicillium* species in foods by attaching antibodies specific for the extracellular polysaccharide of *P. digitatum* to 0.8 μm latex beads. This method has been successful in detecting contamination of grains and processed foods at a limit of detection of 5-10 ng/mL of antigen.

In comparison to other detection assays, the latex agglutination assay exceeds the detection limit of the Enzyme-linked immunosorbent assay (ELISA) and is as effective in detecting *Aspergillus* and *Pencillium* species as the ergosterol production assay. However, the latter displays an increased ability to detect *Fusarium* species when compared to the latex agglutination assay.

8.5. Fungi used in single cell protein production

8.5.1. Candida utilis or Candida tropicalis

Candida utilis **or** ***Torula*** (formerly *Torulopsis utilis, Torula utilis*) is a species of yeast fungi which is used in the production of single cell protein on suitable substrate. These single cell proteins are generally used in animal and poultry feeds. The substrate used in the production of single cell protein are unutilised sugar souces like date fruit syrup or sugarcane bagasse hemicellulose etc.

8.5.2. Fusarium venenatum

*Fusarium venenatum*t has a high protein content. One of its strains is used commercially for the production of the single cell protein mycoprotein.

Fusarium venenatum was discovered growing in Buckinghamshire in the United Kingdom, in 1967 by ICI as part of the effort during the 1960s to find alternative sources of food to fill the protein gap caused by the growing world population. The strain *Fusarium venenatum A3/5* (IMI 145425, ATCC PTA-2684) was developed commercially by an ICI and Rank Hovis McDougall joint venture to derive a mycoprotein used as a food. Because the hyphae of the fungus are similar in length and width to animal muscle fibres, the mycoprotein is used as an alternative to meat and is marketed to vegetarians as Quorn. It is also suitable as a substitute for fat in dairy products and a substitute for cereal in breakfast and snacks.

8.5.3. Kluyveromyces marxianus

Kluyveromyces marxianus (Fig.206) is a species of yeast in the genus *Kluyveromyces*, and is the sexual form (teleomorph) of *Candida kefyr* which is used in the production of single cell protein.

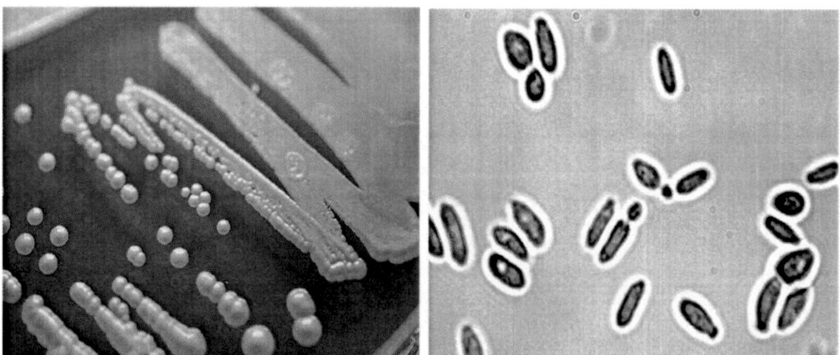

Fig. 206: *Kluyveromyces marxianus* colonies on growth medium and structure of yeast cells

Uses

It is produced as a nutritional yeast and a bonding agent for fodder and pet food, and as a source of ribonucleic acid in pharmaceuticals.

K. marxianus is also used commercially to produce the lactase enzyme similar to the use of other fungi such as those in the genus *Aspergillus*.

8.6. Fungi as a flavouring agents

Candida utilis

Candida utilis or *Torula* (Fig.207), in its inactive form (usually labeled as **torula yeast**), is widely used as a flavouring agent in processed foods and pet foods.

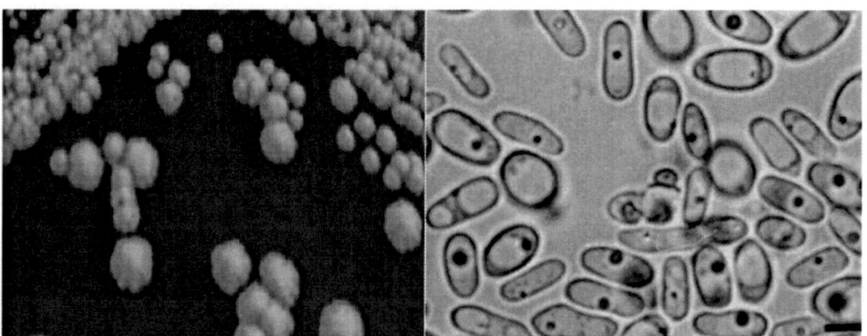

Fig. 207: *Candida utilis* colonies on growth medium and structure of yeast cells

It is often grown on wood liquor, a byproduct of paper production, which is rich in wood sugars. It is pasteurized and spray-dried to produce a fine, light grayish-brown powder with a slightly yeasty odor and gentle, slightly meaty taste.

Torula yeast has become a popular replacement for the flavor enhancer monosodium glutamate (MSG) among manufacturers marketing "all-natural" products.

8.7. Fungi as an organic pest attractant

Candida utilis

Candida utilis or *Torula* finds accepted use in Europe and California for the organic control of olive flies. When dissolved in water, it serves as a food attractant, with or without additional pheremone lures, in McPhail and OLIPE traps, which drown the insects. In field trials in Sonoma County, California, mass trappings reduced damage to an average of 30% compared to almost 90% in untreated controls.

8.8. Fungi useful in enzyme production

8.8.1. Aspergillus wentii (Fig.208)

It is used in the production of enzyme pectinase

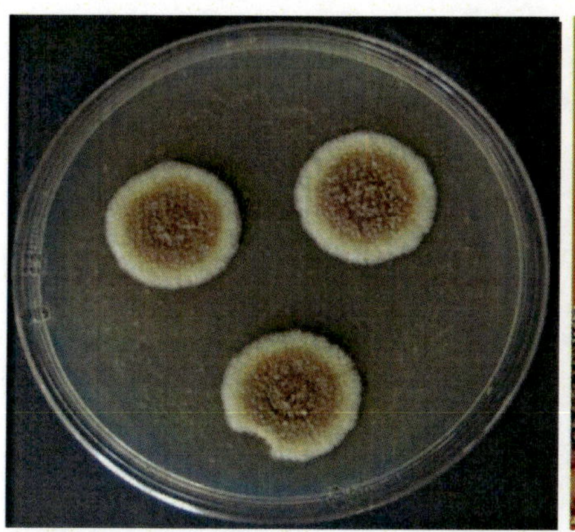

Fig. 208: Fungal colonies on growth media and structure of *Aspergillus wentii*

8.8.2. Aspergillus oryzae

Aspergillus oryzae (Fig.209) is used in production of enzyme amylase which is used in some toothpastes and digestive aid.

It is also used in production of takadiastase which is used for starch and protein splitting.

It is also used in production of enzyme protease.

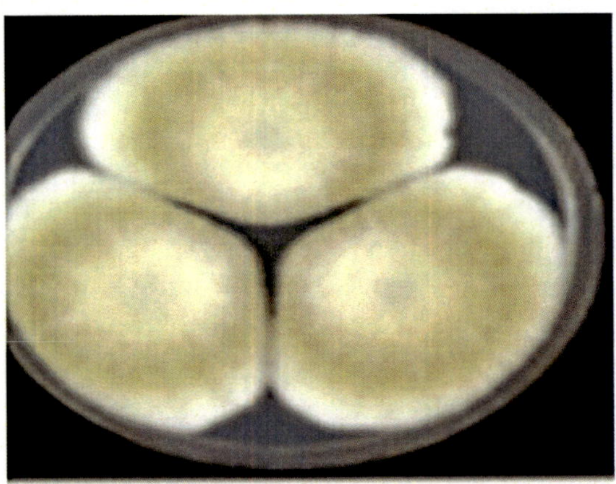

Fig. 209: Fungal colonies of *Aspergillus oryzae* on growth media

8.8.3. Rhizomucor miehei

Rhizomucor miehei (Fig.210) is commercially used to produce enzymes which can be used to produce a microbial rennet to curd milk and produce cheese.

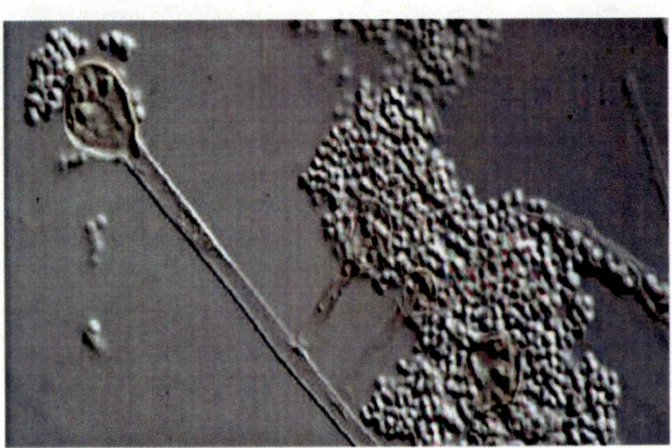

Fig. 210: Fungal structure of *Rhizomucor miehel*

8.8.4. Aspergillus niger

Aspergillus niger (Fig.211) is used in production of enzyme Catalase which is used for Contact lens cleaning systems

It is also used in the production of enzyme Glucose oxidase which is used for Analysis of blood glucose levels and Removal of O_2 from beer, aiding flavor stability and haze life

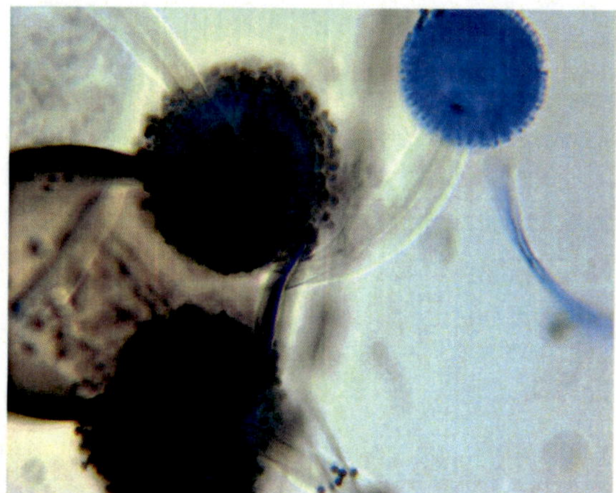

Fig. 211: Spore bearing fruiting bodies of *Aspergillus niger*

8.8.5. Kluyveromyces species

It is used in the production of enzyme Invertase which invert sugar (glucose + fructose).

8.8.6. Kluyveromyces lactis

Kluyveromyces lactis (formerly *Saccharomyces lactis*) has the ability to assimilate lactose and convert it into lactic acid.

K. lactis is also used to produce chymosin (rennet) on a commercial scale.

8.8.7. Trichoderma viride

Trichoderma viride (Fig.212) is used in the production of enzyme cellulase.

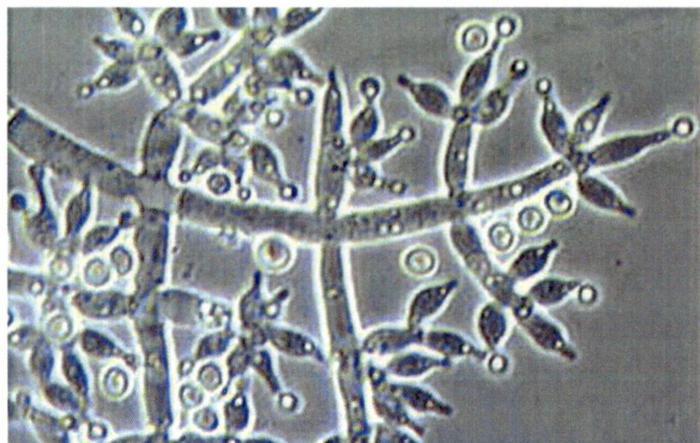

Fig. 212: Fungal structure of *Trichoderma viride*

8.8.8. Candida cylindraceae

It is used in the production of enzyme Lipase

8.8.9. Trichoderma species

It is used in production of enzyme Rhodanase which is used in treatment for cyanide poisoning.

8.9. Fungi used in vitamin production

8.9.1. Blakeslea trispora

Blakeslea trispora (Fig.213) is a source of commercial beta carotene for dietary supplements and food additives.

Beta-carotene is found in many foods and is sold as a dietary supplement. Medical authorities generally recommend getting beta-carotene from food rather than supplementation. There is insufficient research to determine whether there is some minimum level of beta-carotene consumption that is necessary for human health and to identify what problems might arise from insufficient beta-carotene intake.

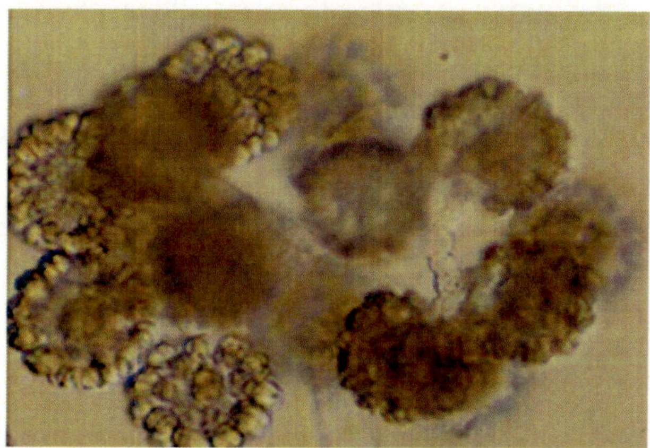

Fig. 213: Fungal spore structure of *Blakeslea trispora*

8.9.2. Ashbya gossypii

Ashbya gossypii (Fig.214) is known to produce Riboflavin. Riboflavin has been used in several clinical and therapeutic situations. For over 30 years, riboflavin supplements have been used as part of the phototherapy treatment of neonatal jaundice.

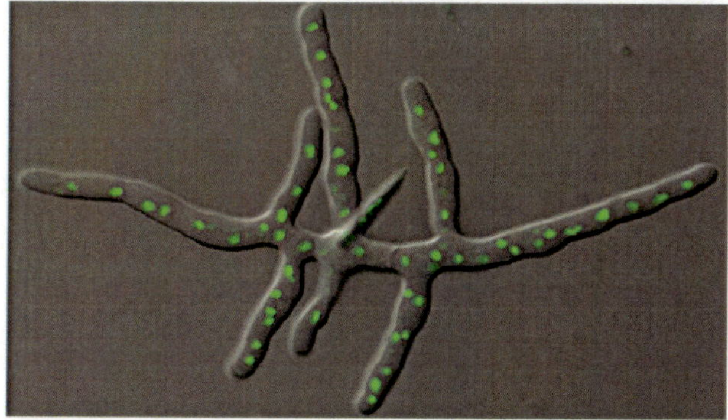

Fig. 214: Fungal structure of *Ashbya gossypii*

8.10. Fungi used in production of polymers

8.10.1. Aureobasidium pullulans

Aureobasidium pullulans is an ubiquitous black, yeast-like fungus that can be found in different environments (e.g. soil, water, air and limestone). It is well known as a naturally occurring epiphyte or endophyte of a wide range of plant species (e.g. apple, grape, cucumber, green beans, cabbage) without causing any symptoms of disease.

A. pullulans (Fig.215) is used in production of pullulan polymer. Pullulan is a polysaccharide polymer consisting of maltotriose unit (three glucose units connected by α-1,4 glycosidic bonds) connected to each other by an α-1,6 glycosidic bond. Pullulan is produced from starch by the fungus *Aureobasidium pullulans*.

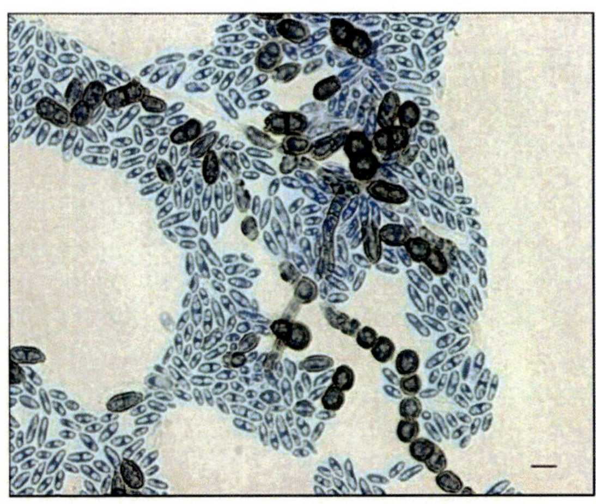

Fig. 215: Fungal spores of *Aureobasidium pullulans*

A. pullulans has also a high importance in biotechnology for the production of different enzymes, siderophores and pullulan.

A. pullulans is also used in biological control of plant diseases, especially storage diseases.

8.10.2. Stolinifer rolfsii

Stolinifer rolfsii is used in production of Scleroglucan polymer.

8.11. Fungi useful in organic acid production

8.11.1. Aspergillus niger

Aspergilus niger is used in production of **citric acid** which is used in soft drinks, candies, artificial lemon juice, baked goods etc.

It is also used in production of Beano which produces an enzyme: alpha-d-galactosidase that suppresses methane production in human digestive tract.

8.11.2. Rhizopus arhizus

Rhizopus arhizus is used for the production of **Fumaric acid** which is used in food production for flavoring and as a preservative.

8.11.3. Aspergillus itaconicus

*It is u*sed in the production of **Itaconic acid**

8.11.4. Aspergillus flavus

It is used in production of **Kojic acid**.

8.12. Fungi useful in production of Fuel and Chemicals

8.12.1. Sacchromyces cerevisiae

It is used in production of **Ethanol**

8.12.2. Zygossachromyces rouxii (Fig.216)

It is used in the production of **Glycerol**.

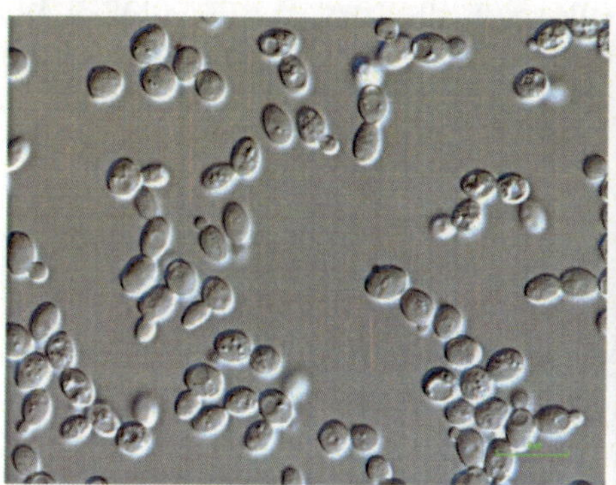

Fig. 216: Yeast cells of *Zygossachromyces rouxii.*

8.13. Fungi used in dye production

Mushrooms can be used to create color dyes via color-extraction with a solvent (often ammonia) as well as particulation of raw material. The shingled hedgehog mushroom and related species contain blue-green pigments, which are used for dyeing wool in Norway. The fruiting body of *Hydnellum peckii* can be used to produce a beige color when no mordant is used, and shades of blue or green depending on the mordant added. *Phaeolus schweinitzii* produces green, yellow, gold, or brown colors, depending on the material dyed and the mordant used.

8.13.1. Cantharellus cibarius

Cantharellus cibarius (Fig.217), commonly known as the **chanterelle**, or **girolle** is probably the best known species of the genus *Cantharellus*. It is used to create dull yellow colour using colour catalyst ammonia.

Fig. 217: Fruiting bodies of *Cantharellus cibarius*

It is also used for culinary purpose. In 1836, the Swedish mycologist Elias Fries considered the chanterelle as one of the most important and best edible mushrooms. Dried chanterelles can also be crushed into flour and used in seasoning in soups or sauces.

8.13.2. Ganoderma applanatum

Ganoderma applanatum (the **Artist's Bracket** or **Artist's Conk**) is a bracket fungus (Fig.218) with a cosmopolitan distribution.This fungus grows as a mycelium within the wood of living and dead trees. It forms fruiting bodies that are up to 30–40 centimetres (12–16 in) across, hard, woody-textured, and inedible; they are white at first but soon turn dark red-brown. Brown spores are released from the pores on the underside of the fruiting body. The fruiting bodies are perennial, and may persist for multiple years, increasing in size and forming new layers of pores as they grow

Fig. 218: Fruiting bodies of *Ganoderma applanatum*

It is used to create rust colour using colour catalyst ammonia.

8.13.3. Agaricus arvensis

Agaricus arvensis (Fig.219), commonly known as the **horse mushroom**, is a mushroom of the genus *Agaricus*.

It is used to create yellowish green colour using colour catalyst salt water.

Fig. 219: Fruiting body of *Agaricus arvensis*

8.13.4. Agaricus campestris

Agaricus campestris (Fig. 220) is a widely eaten gilled mushroom closely related to the cultivated button mushroom *Agaricus bisporus*. It is commonly known as the **field mushroom** or, in North America, **meadow mushroom**.

Fig. 220: Fruiting body of *Agaricus* campestris

It is used to create yellowish green colour using colour catalyst salt water.

Slices of *A. campestris* are applied to scalds, and burns in parts of Scotland. Research into fungal dressings for the treatment of ulcers, and bed sores, using fungal mycelial filaments, is ongoing.

8.13.5. Coprinus comatus

Coprinus comatus (Fig.221), the **shaggy ink cap**, **lawyer's wig**, or **shaggy mane**, is a common fungus often seen growing on lawns, along gravel roads and waste areas. The young fruit bodies first appear as white cylinders emerging from the ground, then the bell-shaped caps open out. The caps are white, and covered with scales.

Fig. 221: Fruiting bodies of the fungus *Coprinus comatus*

It is used to create grayish-green colour using colour catalyst iron pot/ammonia.

8.13.6. *Boletus edulis*

Boletus edulis (Fig.222) (English: **cep**, **porcino** or **porcini**) as its name implies, is an edible mushroom. Italian chef and restaurateur Antonio Carluccio has described it as representing "the wild mushroom *par excellence*", and hails it as the most rewarding of all fungi in the kitchen for its taste and versatility.

Fig. 222. Fruiting bodies of fungus *Boletus edulis*

It is used to create reddish- yellow colour using colour catalyst ammonia.

Boletus edulis is well suited to drying, in which its flavour intensifies. It is easily reconstituted, and its resulting texture is pleasant. Dried porcini have more protein than most other commonly consumed vegetables apart from soybeans.

8.13.7. *Pleurotus ostreatus*

Pleurotus ostreatus, the oyster mushroom (Fig.223), is a common edible mushroom. . It was first cultivated in Germany as a subsistence measure during World War I and is now grown commercially around the world for food. It is related to the similarly cultivated king oyster mushroom. Oyster mushrooms can also be used industrially for mycoremediation purposes.

It is used to create grayish-green colour using colour catalyst iron pot/ammonia.

Fig. 223: Fruiting bodies of the fungus *Pleurotus ostreatus*

As a Culinary uses the oyster mushroom is frequently used in Japanese, Korean and Chinese cookery as a delicacy. Oyster mushrooms are sometimes made into a sauce, used in Asian cooking, which is similar to oyster sauce.

One preliminary study showed that consumption of oyster mushroom extracts lowered cholesterol levels, an effect linked to their content of beta-glucans.

8.13.8. Hypomyces lactifluorum

The **Lobster mushroom**, *Hypomyces lactifluorum* (Fig.224), contrary to its common name, is not a mushroom, but rather a parasitic ascomycete fungus that grows on certain species of mushrooms, turning them a reddish orange color that resembles the outer shell of a cooked lobster.

It is used to create cinnamon pink to red colour using colour catalyst ammonia.

Fig. 224: Fruiting bodies of fungus *Hypomyces lactifluorum*

Lobster mushrooms are widely eaten and enjoyed; they are commercially marketed and are commonly found in some large grocery stores. They have a seafood-like flavor and a firm, dense texture. According to some, they may taste somewhat spicy if the host mushroom is an acrid *Lactarius*.

8.13.9. Phaeolus schweinitzii

Phaeolus schweinitzii (Fig.225), commonly known as **velvet-top fungus**, **dyer's polypore**, or **dyer's mazegill**, is a fungal plant pathogen that causes butt rot on conifers such as Douglas-fir, spruce, fir, hemlock, pine, and larch.

Fig. 225: Fruiting bodies of fungus *Phaeolus schweinitzii*.

It is an excellent natural source of green, yellow, gold, or brown dye, depending on the material dyed and the mordant used. It is use to create colour orange, deep green, rust red, yellow using colour catalyst ammonia, copper pot/ammonia, iron pot/ammonia, and salt water respectively.

8.13.10. Grifola frondosa

Grifola frondosa (Fig.226) is a polypore mushroom that grows in clusters at the base of trees, particularly oaks. The mushroom is commonly known among English speakers as **hen-of-the-woods**, **ram's head** and **sheep's head**. The mushroom is known by its Japanese name ***maitake*** which means "dancing mushroom". Throughout Italian American communities in the northeastern United States, it is commonly known as the **signorina** mushroom.

It is used to create colour light yellow using colour catalyst ammonia

Fig. 226: Fruiting bodies of the fungus *Grifola frondosa*

It is also used in traditional Chinese and Japanese medicine to enhance the immune system.

Researchers have also indicated that whole maitake has the ability to regulate blood pressure, glucose, insulin, and both serum and liver lipids, such as cholesterol, triglycerides, and phospholipids, and may also be useful for weight loss.

Maitake is rich in minerals (such as potassium, calcium, and magnesium), various vitamins (B_2, D_2 and niacin), fibers and amino acids. One active constituent in Maitake for enhancing the immune activity was identified in the late 1980s as a protein-bound beta-glucan compound.

In 2009, a phase I/II human trial, conducted by Memorial Sloan–Kettering Cancer Center, showed Maitake could stimulate the immune systems of breast cancer patients. Small experiments with human cancer patients have shown Maitake can stimulate immune system cells, like NK cells. *In vitro* research has also shown Maitake can stimulate immune system cells. An *in vivo* experiment showed that Maitake could stimulate both the innate immune system and adaptive immune system.

In vitro research has shown Maitake can induce apoptosis in various cancer cell lines as well as inhibit the growth of various types of cancer cells. Small studies with human cancer patients revealed that a portion of the Maitake mushroom, known as the "Maitake D-fraction", possesses anti-cancer activity. *In vitro* research demonstrated the mushroom has potential anti-metastatic properties. Research has shown Maitake has a hypoglycemic effect, and may be beneficial for the management of diabetes. The reason Maitake lowers blood sugar is because the mushroom naturally contains an alpha glucosidase inhibitor.

Maitake contains antioxidants and may partially inhibit the enzyme cyclooxygenase. An experiment showed that an extract of Maitake inhibited angiogenesis via inhibition of the vascular endothelial growth factor (VEGF).

Lys-N is a unique protease found in Maitake. Lys-N is used for proteomics experiments due to its protein cleavage specificity.

8.13.11. Laetiporus

Laetiporus (Fig.227) is a genus of edible mushrooms found throughout the world. Some species, especially *Laetiporus sulphureus,* are commonly known as **sulphur shelf**, **chicken of the woods**, the **chicken mushroom**, or the **chicken fungus** because many think they taste like chicken. The name "chicken of the woods" is not to be confused with the edible polypore, *Maitake* (*Grifola frondosa*) known as "hen of the woods", or with *Lyophyllum decastes*, known as the "fried chicken mushroom".

It is used to create colour orange using colour catalyst ammonia.

Fig. 227: Fruiting bodies of the fungus *Laetiporus*

It can also be used as a substitute for chicken in a vegetarian diet. *Laetiporus sulphureus* has potent ability to inhibit staph bacteria (*Staphylococcus aureus*), as well as moderate ability to inhibit the growth of *Bacillus subtilis*.

8.13.12. Calvatia gigantea

Calvatia gigantea, commonly known as the **Giant puffball** (Fig.228), is a puffball mushroom commonly found in meadows, fields, and deciduous forests usually in late summer and autumn. It is found in temperate areas throughout the world.

It is used to create colour dark red using colour catalyst ammonia.

Fig. 228: Fruiting body of fungus *Calvatia gigantea*

8.13.13. Ganoderma lucidum

The lingzhi mushroom or reishi mushroom is a species complex that encompasses several fungal species of the genus *Ganoderma,* most commonly the closely related species *Ganoderma lucidum* (Fig.229), *Ganoderma tsugae,* and *Ganoderma sichuanense*. *G. sichuanense* enjoys special veneration in East Asia, where it has been used as a medicinal mushroom in traditional Chinese medicine for more than 2,000 years, making it one of the oldest mushrooms known to have been used medicinally. Lingzhi is listed in the *American Herbal Pharmacopoeia and Therapeutic Compendium.*

It is used to create colour orange using colour catalyst ammonia.

Fig. 229: Fruiting boby of the fungus *Ganoderma lucidum*

It has Anti-Allergic/Anti-Inflammatory Activity. Studies showed that Reishi extract significantly inhibited all four types of allergic reactions, including positive effects against asthma and contact dermatitis and effectively used in treating stiff necks, stiff shoulders, conjunctivitis (inflammation of the fine membrane lining the eye and eyelids), bronchitis, rheumatism and improving "competence" of the immune system without any significant side-effects.

It has Anticonvulsant Effects. A water extract from Reishi mycelium significantly increased the threshold for psychomotor seizures in mice. Mycelial extracts also confer anti-inflammatory activity as evidenced by inhibitory activity of lipopolysaccharide (LPS)-induced nitric oxide (NO) production in murine macrophage-like cell line RAW264.7 cells.

It is use in Cancer. The use of *G. lucidum* has also been explored as a complementary adjunct treatment in patients undergoing chemotherapy treatment. A recent meta-analysis of five randomized control trials showed that patients responded more positively when given *G. lucidum* alongside their chemotherapy regimen, and the studies also showed that patients had improved immune functions that was measured by their elevated levels of immune response cells. Several compounds in *G. lucidum* have been studied for apoptotic activity in colon cancer cells, anti-proliferative effects in ovarian cancer cells, and induction of apoptosis in human gastric carcinoma cells.

Cardiovascular Risk Factors. Previous clinical evidence suggested that *G. lucidum* may have antioxidant, cardio-protective, and glycemic regulatory effects. However, a 2015 Cochrane review did not find evidence to support the use of or treatment of cardiovascular risk factors in people with type 2 diabetes mellitus.

Diabetes: Several compounds in *G. lucidum* (including polysaccharides, proteoglycans, proteins and triterpenoids) may have hypoglycemic effects. In vitro evidence suggests that protein tyrosine phosphatase 1B is a promising therapeutic target in diabetes, and a *G. lucidum* proteoglycan can inhibit this enzyme. Secondly, *G. lucidum* demonstrates inhibition of aldose reductase and α-glucosidase, which can suppress postprandial hyperglycemia. A proteoglycan enhanced insulin secretion and decreasing hepatic glucose output (along with increased adipose and skeletal muscle glucose disposal) and normalized serum lipids in a murine model of diabetes. A polysaccharide also demonstrated hypoglycemic effects in type 2 diabetic mice.

Gastrointestinal Health: Recent murine studies suggest that *G. lucidum* may positively impact gut microflora to attenuate metabolic risk factors contributing to obesity.

***Hepatoprotection*:** *G. lucidum* significantly decreased serum ALT and AST levels in mice livers injured with α-amanitin. A proteoglycan also demonstrated hepatoprotective effects in carbon tetrachloride-induced liver injury in vitro and in vivo.

***Immunostimulation*:** *G. lucidum* contains beta glucans and other polysaccharides to stimulate innate immune function and signaling and activate dendritic cells.

***Neuroprotection*:** *G. lucidum* protected dopaminergic neurons through inhibition of microglia.

8.13.14. Blewit

Blewit refers to two closely related species of edible agarics in the genus *Clitocybe*, the **wood blewit** (*Clitocybe nuda*) and the **field blewit** or **blue-leg** (*C.saeva*). Both species are treated by some authorities as belonging to the genus *Lepista*.

Both wood blewits and field blewits (Fig.230) are generally regarded as edible, but they are known to cause allergic reactions in sensitive individuals. This is particularly likely if the mushroom is consumed raw, though allergic reactions are known even from cooked blewits. Wood blewits contain the sugar trehalose, which is edible for most people.

Fig. 230: Fruiting bodies of the fungus Blewit

Field blewits are often infested with fly larvae and do not store very well; they should therefore be used soon after picking. They are also very porous, so they are best picked on a dry day. In most mycologists' opinion, the blewits are considered excellent mushrooms, despite their coloration. Blewits can be eaten as a cream sauce or sautéed in butter, but it is important not to eat them raw, which could lead to indigestion. They can also be cooked like tripe or as omelette filling, and wood blewits also make good stewing mushrooms. It is used to create colour green using colour catalyst ammonia.

9

Fungi as Biocontrol Agent

9.1. Fungi useful against plant disease pathogen

9.1.1. Trichoderma viride

Trichoderma viride (Fig.231) fungus is used as a bio-fungicide. It is used for seed and soil treatment for suppression of various diseases caused by fungal pathogens. It is also a pathogen in its own right,causing green mould rot of onion.

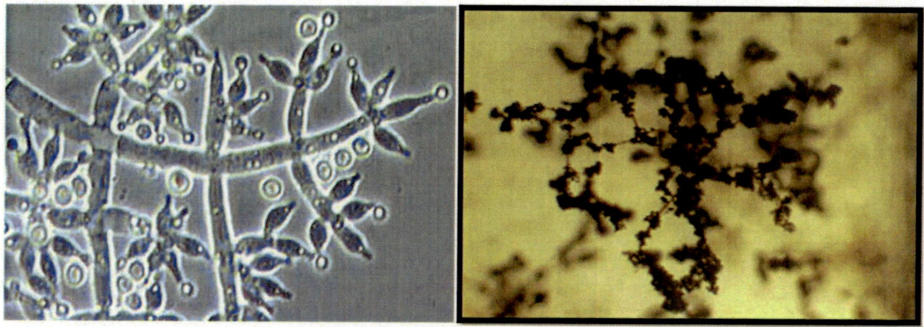

Fig. 231: Fungal structure of *Trichoderma viride*

The fungicidal activity makes *T. viride* useful as a biological control agent against plant pathogenic fungi. It has been shown to provide protection against pathogens like *Rhizoctonia*, *Pythium* and *Armillaria.*

It is found naturally in soil and is effective as a seed dressing bioinoculant in the control of seed and soil-borne diseases including *Rhizoctonia solani*, *Macrophomina phaseolina* and *Fusarium* species. When it is applied to the seed, it colonizes the seed surface and kills not only the pathogens present on the cuticle, but also provides protection against soil-borne pathogens.

A closely related species, *Trichoderma reesei*, is used in the creation of stonewashed jeans. The cellulase produced by the fungus partially degrade the cotton material in places, making it soft and causing the jeans to look as if they had been washed using stones.

9.1.2. Trichoderma herzianum

Trichoderma harzianum fungus is also used as a bio-fungicide. It is used for foliar application, seed treatment and soil treatment for suppression of various disease causing fungal pathogens.*Trichoderma* spp. are present in nearly all soils. In soil, they frequently are the most prevalent culturable fungi. They also exist in many other diverse habitats.

Trichoderma readily colonizes plant roots and some strains are rhizosphere competent i.e. able to grow on roots as they develop. *Trichoderma spp.* also attack, parasitize and otherwise gain nutrition from other fungi. They have evolved numerous mechanisms for both attack on other fungi and for enhancing plant and root growth. Different strains of *Trichoderma* control almost every pathogenic fungus for which control has been sought. However, most *Trichoderma* strains are more efficient for control of some pathogens than others, and may be largely ineffective against some fungi.

9.1.3. Gliocladium virens

Gliocladium virens (Fig.232) is a naturally occurring, ubiquitous soil saprophyte and has been shown to suppress a variety of soilborne plant pathogens, including *Pythium* spp. *Rhizoctonia solani*, and *Sclerotium rolfsii* that cause damping-off, root rots, and various other seedling diseases on a wide variety of host plants.

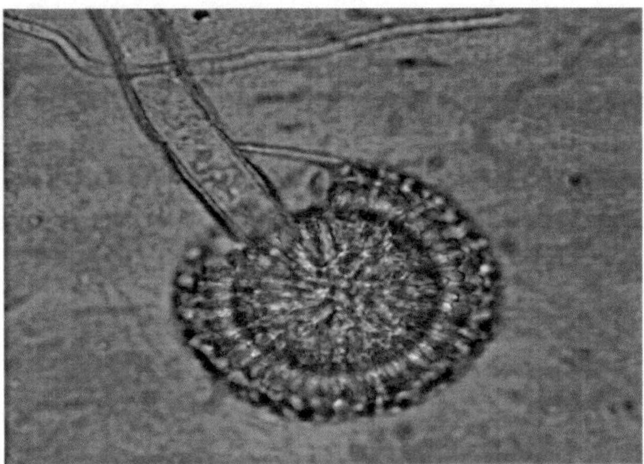

Fig. 232: Fungal structure of *Gliocladium virens*

9.1.4. Ampelomyces quiqualis

The fungus *A. quisqualis* (Fig.233) is reported to be a hyperparasite of powdery mildew where it can be easily found associated with powdery mildew colonies. Hyphae of *Ampelomyces* penetrate the hyphae of powdery mildews and grow internally to kill all the parasitized cells. *Ampelomyces quisqualis* is the mycoparasitic anamorphic ascomycete that reduces the growth and kills powdery mildews pathogen.

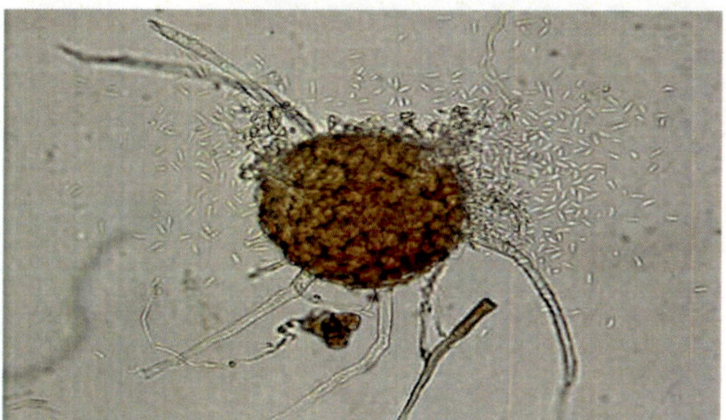

Fig. 233: Fungal structure of *Ampelomyces quiqualis*

It can affect the pathogen through antibiosis and parasitism.

9.1.5. Chaetomium globosum and Chaetomium cupreum

The fungus has biocontrol activity against root rot disesease caused by *Fusarium, Phytophthora* and *Pythium.*

Ascospore inoculation of chaetomium (Fig.234) reduces bacterial disease symptoms such as wilting, and fungal disease like apple scabs, and seed blight in treated plants.

It enhances plant stress tolerance and microbial defense, renders *C. globosum* application beneficial for agricultural use. The effect of heavy metals such as copper which suppress plant growth and disrupt metabolic processes, e.g. photosynthesis is minimize when maize plants were treated with *C. globosum*, where they expressed less growth inhibition and increased biomass. *C. globosum* is also known to reside in Ginkgo biloba plants.

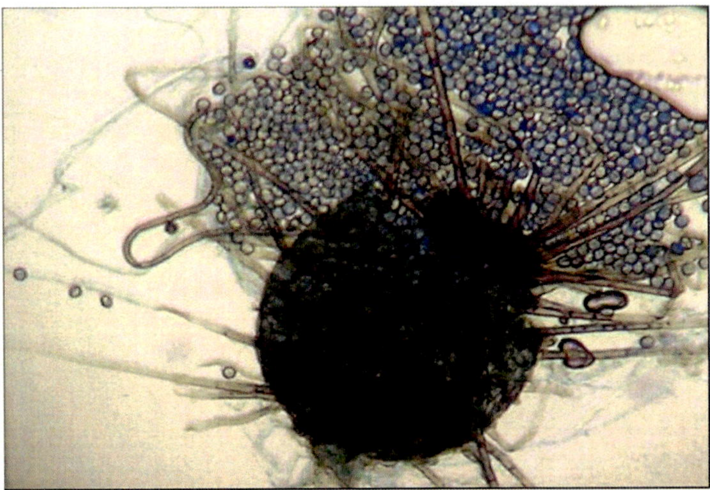

Fig. 234: Fungal structure of *Chaetomium*

9.1.6. Coniothyrium minitans

Coniothyrium minitans (Fig.235) is reported to be a mycoparasite of *Sclerotinia* species such as *Sclerotinia minor, S. sclerotiorum, S. trifoliorum* and *S. cepivorum*

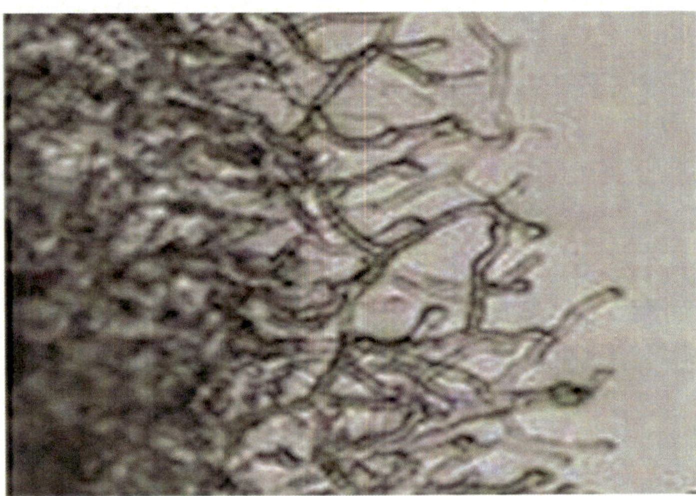

Fig. 235: Fungal structure of *Coniothyrium minitans*

It has been applied successfully to control disease in many crops including lettuce, oilseed rape, peanut and alfalfa.

9.1.7. Pythium oligandrum

Pythium oligandrum has shown ability to control soil-borne pathogens both in the laboratory and in the field. *Pythium oligandrum* oospores (Fig.236) have been applied as seed treatments which reduce damping-off disease caused by *P. ultimum* in sugarbeet

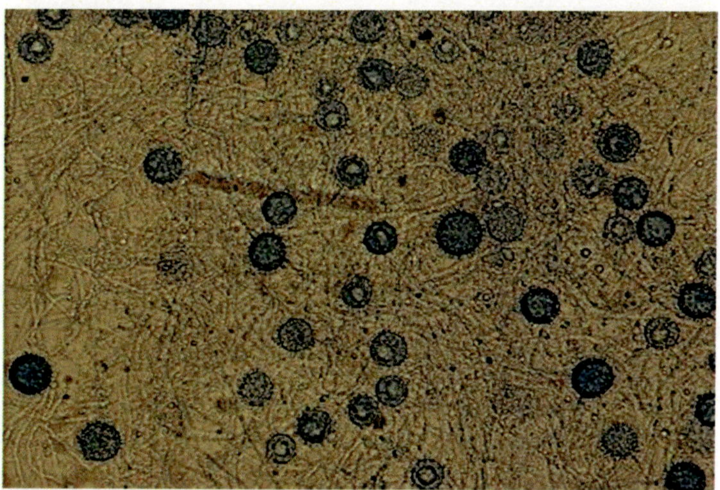

Fig. 236: Oospores of *Pythium oligandrum*

9.1.8. Lactarious lacta (Ectomycorrhizal fungi)

Lactarious lacta strongly suppressed *Fusarium oxysporum* under nursery condition.

9.2. Fungi useful against insect pest of crop plants

9.2.1. Verticillium lecanii

Verticillium lecanii (Fig.237) is an entomopathogenic fungus. The mycelium of this fungus produces a cyclodepsipeptide toxin called bassianolide and other insecticidal toxins such as dipicolinic acid, which infect aphids, whiteflies, scale insects and lead to their death.

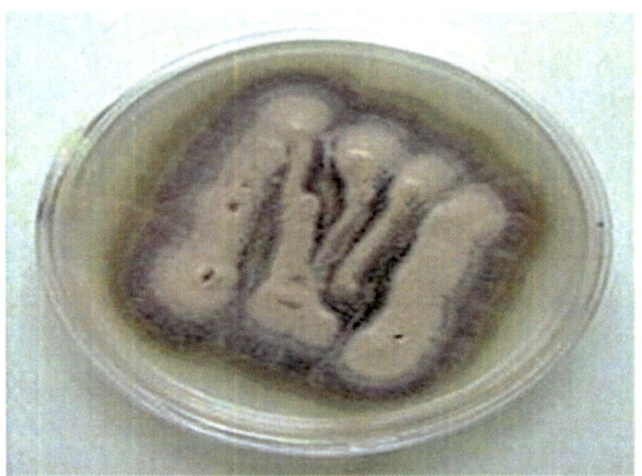

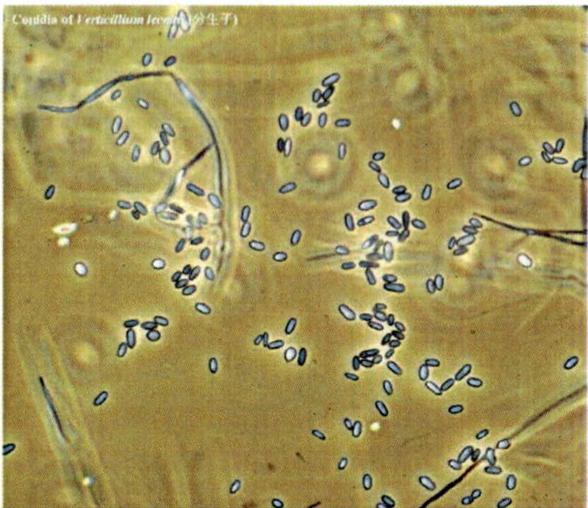

Fig. 237: Fungal growth of Verticillium lecanii on growth medium and the spores of the fungus

It is used against whitefly, thrips and aphids

The product develop from *Verticillium lecanii* are Mycotal and Vertalec and producers of these product are Koppert in Netherlands.

9.2.2. Beauveria bassiana

The fungus *Beauveria bassiana* grows naturally in soils throughout the world and acts as a parasite on various arthropod species, causing **white muscardine disease in the insect.**

It is being used as a biological insecticide to control a number of pests such as **termites, thrips, whiteflies, aphids and different beetles**. Its use in the control of bedbugs and trails for malaria-transmitting mosquitos is under investigation.

As a species, *Beauveria bassiana* (Fig.238) parasitizes a very wide range of arthropod hosts. However, different strains vary in their host ranges, some having rather narrow ranges, like strain Bba 5653 that is very virulent to the larvae of the Diamond back Moth and kills only few other types of caterpillars.

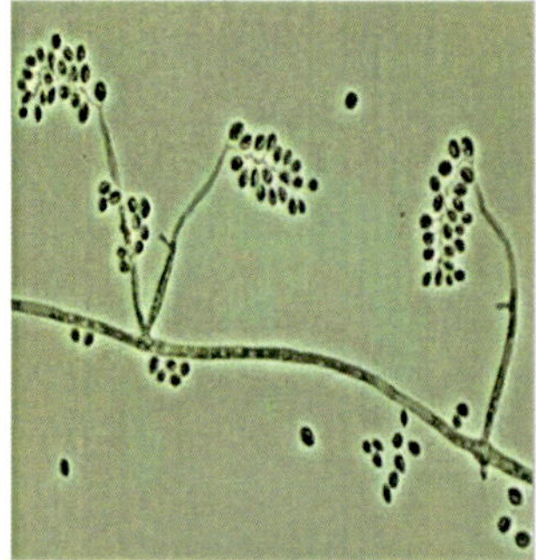

Fig. 238: Fungal structure of *Beauveria bassiana* and the parasitiosed bugs

Preliminary research has shown that the fungus is 100% effective in eliminating bed bugs exposed to cotton fabric sprayed with fungus spores.The fungus is carried by the infected bugs back to their harborages where it affect other bed bug colonies and all colonies die within 5 days of exposure.

It is used to control pests like coffee berry borer, corn borer, European corn borer, Army worm and Cotton pest.

The products develop from B. bassiana are Mycotrol GH and Mycotrol WP and producers of these product is Mycotrol in USA

9.2.3. Metarhizium anisopliae

The fungus Metarhizium anisopliae (Fig.239), formerly known as *Entomophthora anisopliae* grows naturally in soils throughout the world and causes disease in various insects by acting as a parasitoid.

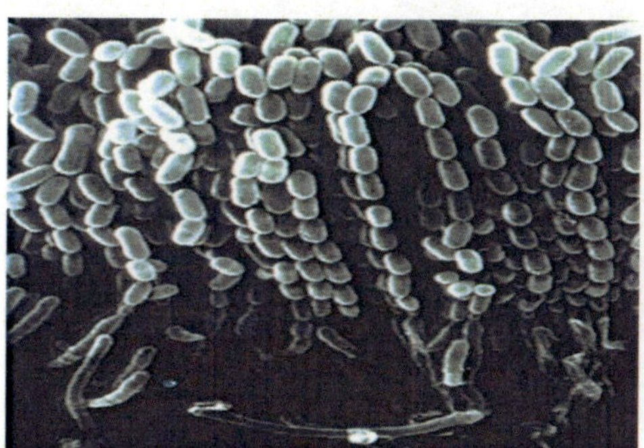

Fig. 239: Fungal structure of *Metarhizium anisopliae* and the parasitiosed insect

M. anisopliae and its related species are used as biological insecticides to control a number of pests such as **Spittlebug, Cockroches, Thrips and Termites,** etc. and its use in the control of malaria-transmitting mosquitoes is under investigation. *M. anisopliae* does not appear to infect humans or other animals and is considered to be safe as an insecticide. The microscopic spores are typically sprayed on affected areas.

Species of *Metarrhizium* infects a number of insects, forming long chains of spores, a feature that has enabled its use in novel roach traps using the fungus rather than chemicals.

The use of a fungus in roach traps is superior to chemical use because chemicals will kill only the insects that enter the chamber; whereas, insects that become infected with *Metarrhizium* will carry the fungus to their hiding places and infect their neighbors.

It is also used to control pest like Vine weevil and Spittle bug.

The product develop from *M. anisopilae* are Metaquino Bio-path and Bio-blast and the producers are Brazil and Eco- science, USA.

9.2.4. Noumorea rileyi

The fungus *Noumorea*, and its species grows naturally in soils throughout the world and causes disease in various insects by acting as a parasitoid. It mainly parasitise the larves of loopers and semiloopers (Fig.240).

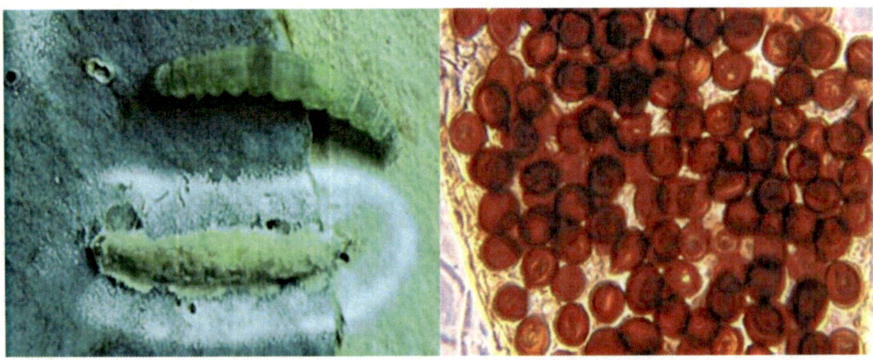

Fig. 240: Parasitization of larves by *Noumorea* and Insect cavity filled with fungal spores

To control the Cabbage loopers, the fungus parasities the larval body cavity which becomes overwhelmed with fungal spores

9.2.5. Cordyceps

The species cordyceps (Fig.241) infect the larvae of many beetles and moths even those deeply embedded in soil.

Fig. 241: *Cordyceps* parasitizing beetle larvae in soil

9.2.6. Stibella

The species zoophora parasities on flies; while the species stibella (Fig.242) parasities on weevils

Fig. 242: A weevil infected with *Stilbella* sp

9.3. Fungi useful against nematode

9.3.1. Dactylella

Dactylella is a genus comprising 72 species of mitosporic fungi (Fig.243) in the family Orbiliaceae. They are notable for trapping and eating nematodes.

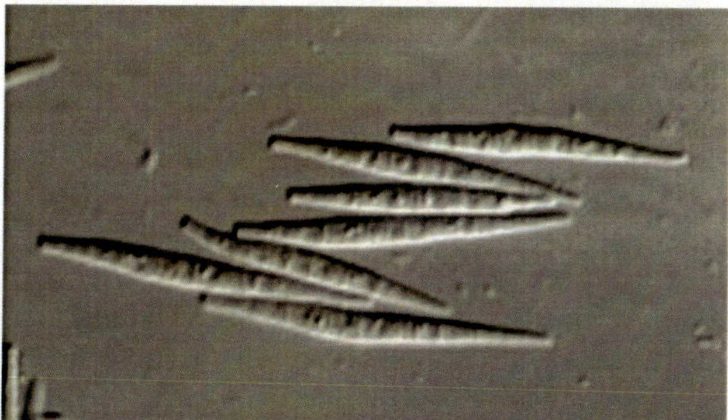

Fig. 243: Fungal spores of *Dactylella* sp.

Members of this genus form a noose structure from several elongate cells. When stimulated by a nematode passing through the structure, the cells swell, tightening the noose and trapping the nematode. Filaments then grow into the nematode to absorb nutrients.

9.3.2. Arthrobotrys

Arthrobotrys is a genus of mitosporic fungi in the family *Orbiliaceae*. There are 71 species. They are predatory fungi that capture and feed on nematode worms (Fig. 244). Rings that form on the hyphae constrict and entrap the worms, then hyphae grow into the worm and digest it.

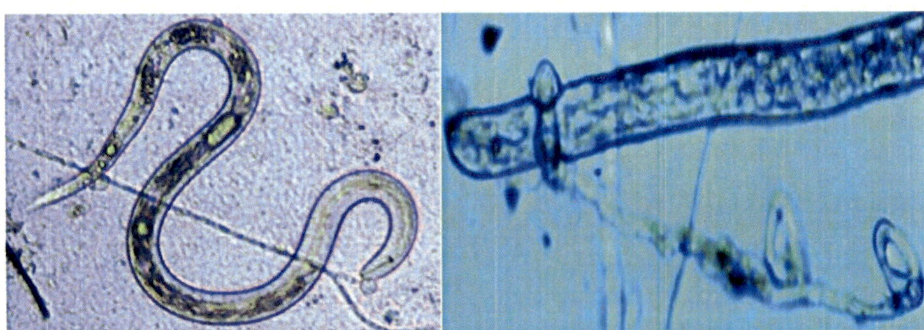

Fig. 244: Formation of a net-like trap by *Arthrobotrys oligospora*

9.3.3. Purpureocillium liliacinum

The fungus (Fig.245) parasitise the eggs of *Meloidogyne incognita* nematode

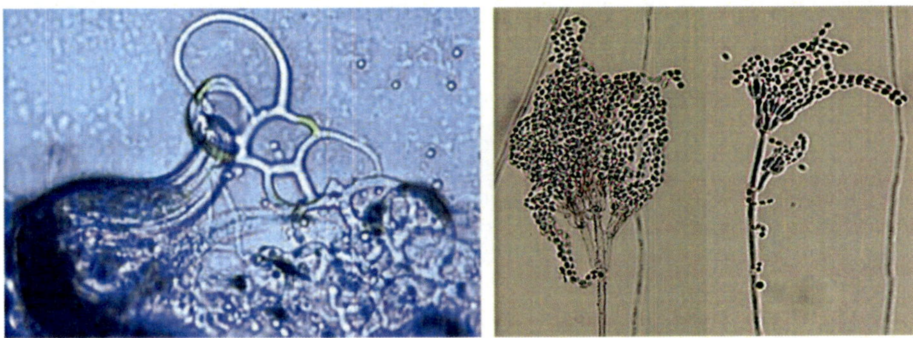

Fig. 245: *Purpureocillium liliacinum* and parasitisation of nematode eggs.

9.4. Fungi useful in the control of noxious weeds

9.4.1. Puccinia obtegens

The rust species *Puccinia obtegens* (Fig.246) has shown some promise for controlling Canada thistle, but it must be used in conjunction with other control measures to be effective.Another leaf rust species, *Puccinia myrsiphylli*, has been used as an effective biocontrol agent for infestations of the common form of bridal creeper (*Asparagus asparagoides*) in Australia since 2000.

Fig. 246. Fungus *Puccinia obtegens* as biocontrol agent against skeleton weed

Phytophthora palmivora has been used to control milkweed or strangler vine, a major problem on citrus in south Florida. The mycoherbicide "Divine" in the form of *P. palmivora* is prepared for the biocontrol of this weed pest by Abbott Labs incorporates.

9.4.2. Colletotrichum gloeosporoides

"Collego" is a relatively new mycoherbicide that utilizes *Colletotrichum gloeosporoides* (Fig.247) to control joint vetch in rice fields, where chemical control presents a real problem. Joint vetch can be a major problem to rice farmers because in the harvesting of rice, the joint vetch seeds, that are similar in size, contaminate the rice and lowers its market value. Several preparations of collego are available

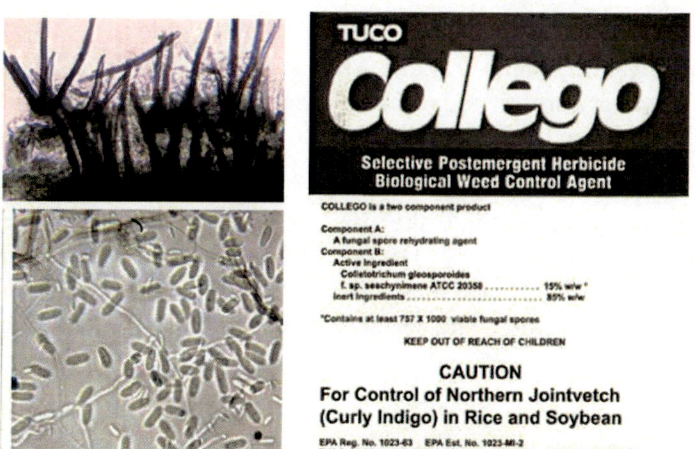

Fig. 247: Fungal structure of *Colletotrichum gloeosporoides* as mycoherbicide

9.4.3 Alternaria cassiae

Alternaria cassiae has been found to be pathogenic to sicklepod plant, a member of pea family, which is a weed in number of crops. In experimental tests it was shown to kill sprayed plants, while unsprayed plants grew vigorously. Similar results were found in field tests.

9.4.4. Cercospora piaropi

Water hyacinth was introduced into Florida in the 1890s. Within the past century, it has spread over almost ½ million acres of waterways. It is a beautiful plant liked by people who put them in the gold-fish ponds, fountains, and other bodies of water from which it escaped to other water bodies.

In 1971, a sharp decline of waterhyacinth was noted in the now famous Rodman Reservoir. It was soon determined that it was highly infected with a species of Cercospora, later described as *C. rodmanii,* renamed as *C. piaropi* (Fig.248) as a biocontrol agent

Fig. 248: Biological control of waterhyacinth by *Cercospora piaropi*

Tests were done on waterhyacinth in Lake Alice at the University of Florida where the results were very promising. Although good biocontrol was achieved, difficulties were experienced in the application of inoculum and the survival ability of waterhyacinth.

Some of the success stories include the use of a new fungus Phomopsis amaranthicola to control of pigweed, Dactylaria higginsii to control purple sedge, Dreschlera/Exerohilum cocktail to control various grasses; Bipolaris saccharii to control cogon grass; and Uromycicola tepperianium to control invading locust trees.

10

Fungi Used as Biofertilizers

10.1. Phosphate Solubilisers

10.1.1. Aspergillus spp.

Several species of *Aspergillus* (Fig.249) have been reported to be involved in the solubilization of inorganic phosphates such as *A. flavus*, *A. fumigates, A. awamori. A. niger* and *A. terreus*.

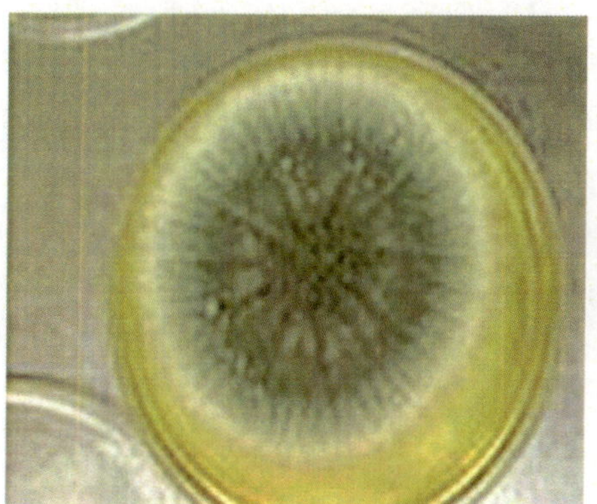
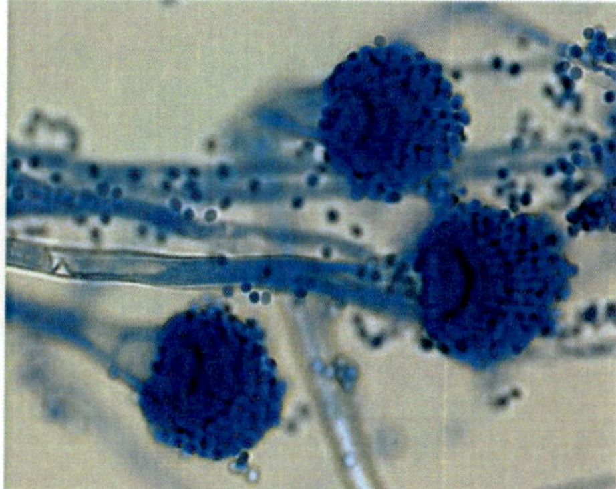

Fig. 249: Growth of *Aspergillus* on media and fungal structure

These fungi are able to solubilize inorganic phosphate through the production of acids viz. citric, gluconic, glycolic, oxalic acids and suçcinic acid.

Aspergillus fumigates isolated from compost has been reported to be potassium releasing fungus.

10.1.2. Penicillium digitatum

Penicillium digitatum (Fig.250) are phosphate solubilizing microorganisms that improve phosphorus absorption in plants and stimulate plant growth.

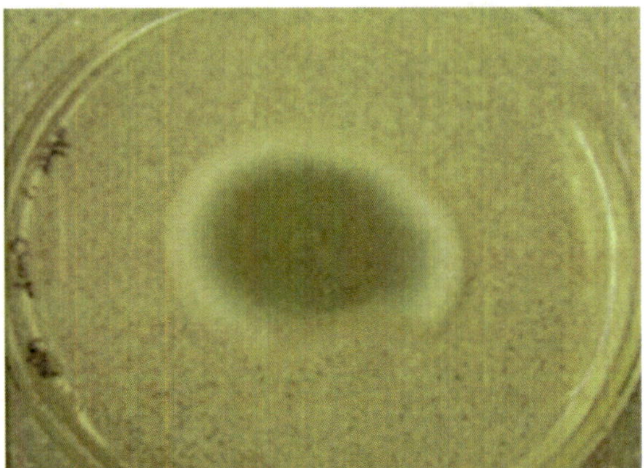

Fig. 250: Growth of *Penicillium digitatum* on media

Similarly, Penicillium bilaiae has been formulated as a commercial product named Jumpstart® and was released to the market as a wettable powder in 1999. *Penicillium bilaiae* is applied to increase dry matter, phosphorus (P) uptake and seed yield in canola (*Brassica napus*).

Penicillium radicum and *P. Italicum* are also phosphate-solubilizing taxa.

Penicillium radicum isolated from the rhizospher of wheat roots, has shown a good promise in plant growth promotion while *P. Italicum* isolated from the rhizosphere soil was tested for its ability to solubilize tri-calcium phosphate (TCP) and could promote the growth of soybean.

10.1.3. Trichoderma viride

Trichoderma species improve mineral uptake, release minerals from soil and organic matter, enhance plant hormone production, induce systematic resistance mechanisms, and induced root systems in hydroponics. For these reasons *Trichoderma* species are known as plant growth promoting fungi.

Trichoderma (Fig.251) species have therefore, successfully been used as biofungicides and biofertilizers in greenhouse and field plant production. There are many *Trichoderma* products as fungal biofertilizers available in the market

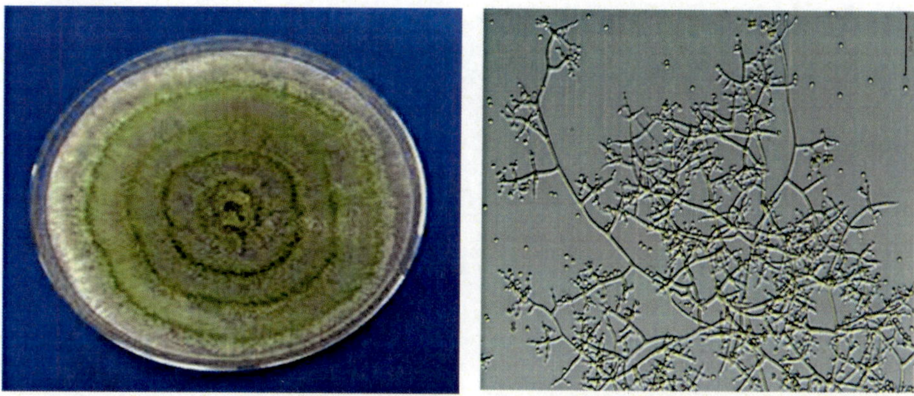

Fig. 251: Growth of *Trichoderma* viride on media and the fungal structure

Trichoderma species not only reduce the occurrence of disease and inhibit pathogen growth when used as mycofungicides, but they also increase the growth and yield of plants. They also increase the survival of seedlings, plant height, leaf area and dry weight.

Trichoderma spp are important sources of antibiotics, enzymes, decomposers and plant growth promoters.

Products

- *Trichoderma harzianum and Trichoderma hamatum*-Bioorganic Plus
- *Trichoderma* spp. - Superzyme and Tricho®

10.1.4. Chaetomium globosum

It is a saprophytic fungus that primarily resides on plants, soil, straw, and dung. They are found in habitats ranging from forest plants to mountain soils across various biomes. *C. globosum* (Fig.252) colonies can also be found indoors and on wooden products.

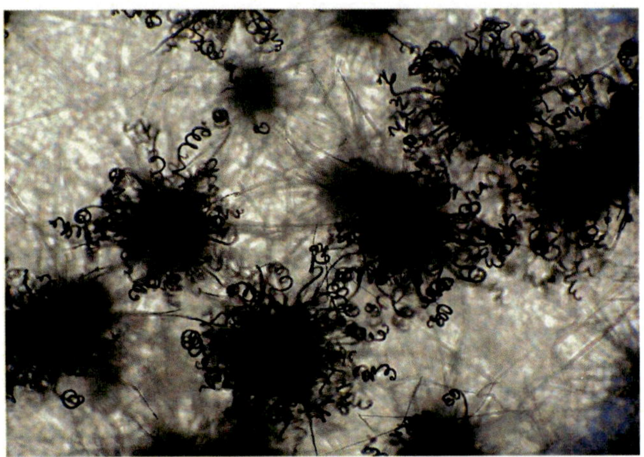

Fig. 252: Fungal structures of *Chatomium globosum*

The product of *Chaetomium* species can be fungal biofertilizers for example Ketomium® which is formulated from *Ch. Globosum* and *Ch. cupreum*is not only a mycofungicide but also plant growth stimulant because tomato, corn, rice, pepper, citrus, durian, bird of paradise and carnation treated with Ketomium® have a greater plant growth and high yields than non-treated plants.

10.2. Mangnese solubilizer

10.2.1. Aspergillus sp *and penicillium* sp.

Manganese solubilizing ability of the fungal isolates is possibly due to the mycelia production of biogenerated organic acids such as oxalic acid, citric acid, maleic acid and gluconic acid and signifies its role in biotransformation of insoluble manganese oxide.

10.3. Vesicular Arbuscular Mycorrhizal Fungi

The vesicular arbuscular Mycorrhizal Fungi (VAM Fungi) are the group of symbiotic, endotrophic-mycorrhizal fungi found in roots of higher plants. They are included in the family *Endogonaceae* of zygomycetes. VAM fungus infects a plant root and forms vesicles and arbuscles in the root cortex and a permanent mantal of hyphae on the roots surface.

VAM fungi help the plant to intake more Zn, S, Cu, P, Ca, K, Fe, Mn and Br from the Soil. It increases the growth rate in plant eg. citrus, maize, wheat, barley.etc. It increases the absorption of water by plants from the soil. It helps to overcome water stress in the soil when drought prevails. VAM fungal infection increase the concentration of cytokinins and chlorophylls in plants. It reduces the sensitivity of crop towards high level of salts and heavy metals in the soil. It improves the hardness of transplant stocks by serving as an extra root hairs,. VAM provides resistance to plants against various soil borne plant pathogen causing root diseases. In fumigated soils plants show stunted growth. VAM fungal reduces the stunting of plants in such soils. VAM fungal infection increases the yield in crops like potato, maize, barley., etc. When the infected plant is starving for food, VAM gives the plant its own food and protect the plant. VAM improved growth, nutrient uptake and yield of cereals, pulses, vegetables (except cruciferous crops), plantation crops, ornamental and forest trees. The crops like *Allium sativum* and *A. cepa* were found to be highly responsive to Mycorrhizal inoculants. Transplanted vegetable crops like tomato, chilli, onion, brinjal.etc find wider scope and feasibility for utilization of VAM inoculants. The VAM inoculums requirements for such crops is minimized by growing the seedlings with mycorrhizal inoculums and such mycorrhizal seedlings when transplanted have been found to show remarkable increase in plant growth nutrient uptake and yields.

Five Genera of VAM fungi (Fig.253) are recognized based of the spore morphology. These are ***Glomus*** (*Glomus microcarpum, G. fasciculatum, G. mosseae* etc.), ***Gigaspora*** (Gigaspora nigra.), ***Acaulospora*** (*Acaulospora scobiculate),* ***Sclerocystis*** (*Sclerocystis clavispora) and* ***Endogone***. (*Endogone increseta)*

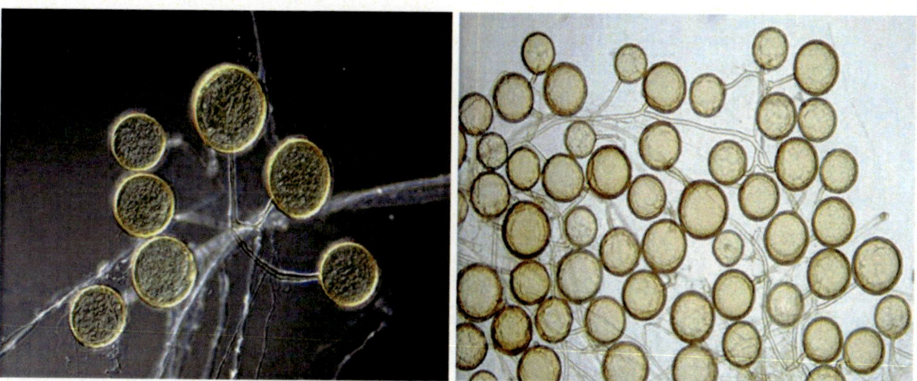

Fig. 253: Spores of VAM fungi *Glomus*

Product: VAM fumgi *Glomus intraradices* is available under the brand name Mycomax.

11

Fungi Used as Decomposer

11.1. *Pleurotus ostreatus*

Researchers in Mexico have shown that oyster mushrooms can break down/ decompose disposable diapers

11.2. *Chaetomium globsum*

It is a saprophytic fungus that primarily resides on plants, soil, straw, and dung. They are found in habitats ranging from forest plants to mountain soils across various biomes. *C. globosum* colonies can also be found indoors and on wooden products

C. globosum assists in cellulose decomposition of plant cells.

11.3. *Aspergillus* spp.

Aspergillus and other moulds play an important role in decomposer consortia because they are adept at recycling starches, hemicelluloses, celluloses, pectins and other sugar polymers. Some aspergilli are capable of degrading more refractory compounds such as fats, oils, chitin, and keratin. Maximum decomposition occurs when there is sufficient nitrogen, phosphorus and other essential inorganic nutrients.

For *Aspergillus* the process of degradation is the means of obtaining nutrients. When these moulds degrade human-made substrates, the process usually is called biodeterioration. Both paper and textiles (cotton, jute, and linen) are particularly vulnerable to *Aspergillus* degradation. Our artistic heritage is also subject to *Aspergillus* assault, for example after Florence (in Italy) flooded in 1969, 74% of the damaged Ghirlandaio fresco in the Ognissanti church were due to *Aspergillus versicolor*.

11.4. *Humicola* spp, *Torula* spp, *Penicillium* spp, *Thermomyces* and *Mucor*

These Fungi are the major organisms decomposing dead leaves and other organic matter. Composting is process for converting decomposable materials into useful stable products. Organic matter is broken down into carbon dioxide and the mineral forms of nutrients like nitrogen. Composting is one of the only ways to revitalize soil vatility due to nutrient depletion in soil.

Treating biodegradable waste before it enters a landfill reduces global warming from fugative methane; untreared waste breaks down anaerobically in a landfill, producing landfill gas methane, a potent greenhouse gas.

Trichoderma harzianum, a biological plant protection agent, is now successfully incorporated into such composts to have additional benefits of soil-borne disease management.

11.5. Brown rot fungi

Brown-rot fungi break down hemicellulose and cellulose. Cellulose is broken down by hydrogen peroxide (H_2O_2) that is produced during the breakdown of hemicelluloses. Because hydrogen peroxide is a small molecule, it can diffuse rapidly through the wood, leading to a decay that is not confined to the direct surroundings of the fungal hyphae. As a result of this type of decay, the wood shrinks, shows a brown discoloration, and cracks into roughly cubical pieces; hence the name *brown rot* or *cubical brown rot.*

Brown-rot fungi of particular economic importance include *Serpula lacrymans* (true dry rot), *Fibroporia vaillantii* (mine fungus), and *Coniophora puteana* (cellar fungus), which may attack timber in buildings. Other brown-rot fungi include the sulfur shelf, *Phaeolus schweinitzii*, and *Fomitopsis pinicola.*

11.6. Soft rot fungi

Soft-rot fungi secrete cellulase from their hyphae, an enzyme that breaks down cellulose in the wood. This leads to the formation of microscopic cavities inside the wood, and sometimes to a discoloration and cracking pattern similar to brown rot. Soft-rot fungi need fixed nitrogen in order to synthesize enzymes, which they obtain either from the wood or from the environment. Examples of soft-rot-causing fungi are *Chaetomium*, *Ceratocystis*, and *Kretzschmaria deusta (Fig.254).*

Fig. 254: Fruiting bodies of soft rot fungi Kretzschmaria

Soft-rot fungi are able to colonise conditions that are too hot, cold or wet for brown or white-rot to inhabit. They can also decompose woods with high levels of compounds that are resistant to biological attack. Bark in woody plants contains a high concentration of tannin, which is difficult for fungi to decompose, and suberin which may act as a microbial barrier. The bark acts as form of protection for the more vulnerable interior of the plant. Soft-rot fungi do not tend to be able to decompose matter as effectively as white-rot fungi: they are less aggressive decomposers.White rots break down lignin and cellulose and commonly cause rotted wood to feel moist, soft, spongy, or stringy and appear white or yellow.

White-rot fungi break down the lignin in wood, leaving the lighter-colored cellulose behind; some of them break down both lignin and cellulose. Because white-rot fungi are able to produce enzymes, such as laccase, needed to break down lignin and other complex organic molecules, they have been investigated for use in mycoremediation applications.

There are many different enzymes that are involved in the decay of wood by white-rot fungi, some of which directly oxidize lignin. The relative abundance of phenylpropane alkyl side chains of lignin characteristically decreases when decayed by white-rot fungi. It has been reported that the oyster mushroom (*Pleurotus ostreatus*) preferentially decays lignin instead of polysaccharides. This is different from some other white-rot fungi e.g. *Phanerochaete chrysosporium*, which shows no selectivity to lignocellulose.

Honey mushroom (*Armillaria spp.*) is a white-rot fungus notorious for attacking living trees. *Pleurotus ostreatus* and other oyster mushrooms are commonly cultivated white-rot fungi, but *P. ostreatus* is not parasitic and will not grow on a living tree, unless it is already dying from other causes. Other white-rot fungi include the turkey tail, artist's conch, and tinder fungus.

White-rot fungi are grown all over the world as a source of food, for example the Shiitake mushroom, which in 2003 comprised approximately 25% of total mushroom production.

12

Fungi Used in Antibiotics Production

12.1. *Penicillium chrysogenum*

Penicillium chrysogenum (Fig.255) (previously known as *Penicillium notatum*), produces the antibiotic penicillin, known as WONDER DRUG.

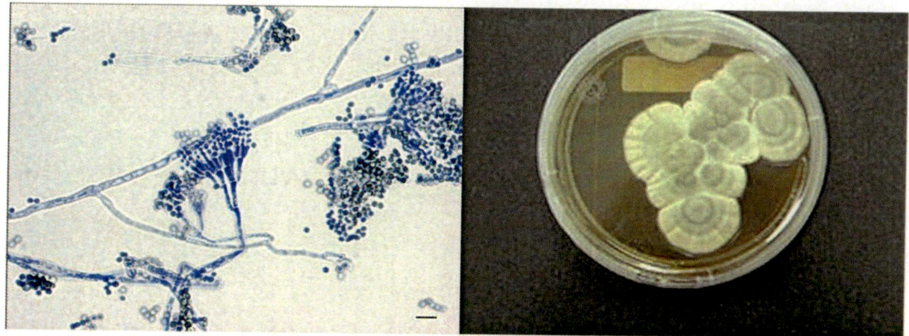

Fig. 255: Fungal structure of *Penicillium chrysogenum* and fungal growth on media

The antibiotics produced from fungus Penicillium is listed in table.

Generic	Brand Name
Amoxicillin	Amoxil, Polymox, Trimox, Wymox
Ampicillin	Omnipen, Polycillin, Polycillin-N, Principen, Totacillin
Bacampicillin	Spectrobid
Carbenicillin	Geocillin, Geopen
Cloxacillin	Cloxapen
Dicloxacillin	Dynapen, Dycill, Pathocil
Flucloxacillin	Flopen, Floxapen, Staphcillin
Mezlocillin	Mezlin
Nafcillin	Nafcil, Nallpen, Unipen
Oxacillin	Bactocill, Prostaphlin
Penicillin G	Bicillin L-A, Crysticillin 300 A.S., Pentids, Permapen, Pfizerpen, Pfizerpen-AS, Wycillin

Penicillin V	Beepen-VK, Betapen-VK, Ledercillin VK, V-Cillin K
Piperacillin	Pipracil
Pivampicillin	
Pivmecillinam	
Ticarcillin	Ticar

Penicillin has also been shown to be produced by a wide range of organisms including the fungi *Aspergillus, Malbranchea, Cephalosporium, Emericellopsis, Paecilomyces, Trichophyton, Anixiopsis, Epidermophyton, Scopulariopsis, Spiroidium* and the actionomycete, *Streptomyces.* The only type of penicillin produced by actinomycetes is Penicillin N (with the chemical structure D- (- aminoadipyl) penicillin usually accompanied by cephamycins and/or deacetyl – 3 – 0-carbamoyl cephalosporin C.

Penicillin is the antibiotic which fights against bacteria in our body. Penicillin is used to treat many different types of infections caused by bacteria, such as pneumonia, scarlet fever and ear, skin and throat infection. It is also used to prevent recurrent rheumatic fever and chorea. It is also used against rickettsia and some of the larger viruses. Penicillin may be used to treat infections such as urinary tract infections, Septiceamia, Meningitis, Intra abdominal infections, Gonorrhoea, Syphilis, Pnemonia, and respiratory infection. Different commercial antibiotic products (Fig.256) are derived from this fungus.

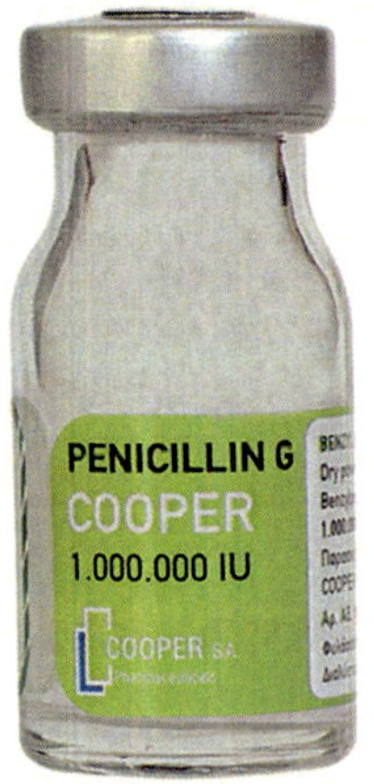

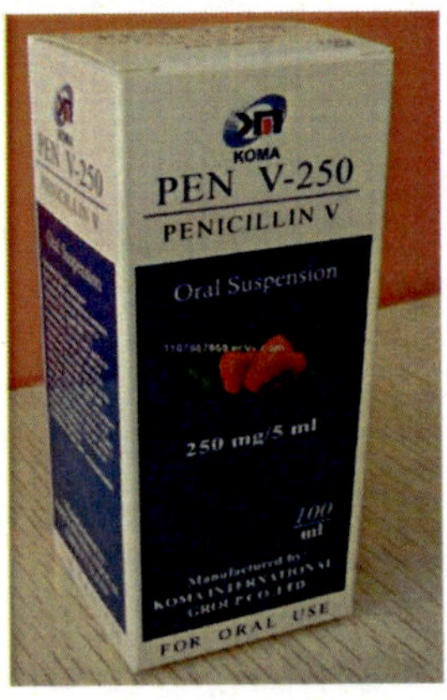

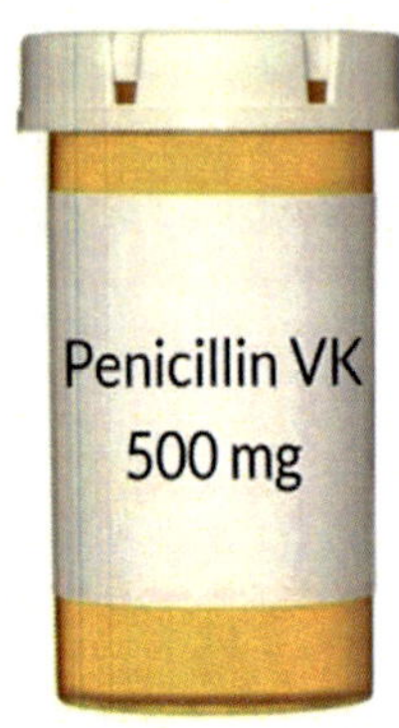

Fig. 256: Commercial products derived from the fungus *Penicillium*

12.2. *Penicillium griseofulvin*

The antibiotic Griseofulvin is produced from the fungus *Penicillium griseofulvin* (Fig.257). Griseofulvin is an antifungal drug that fight against infection caused by fungus.

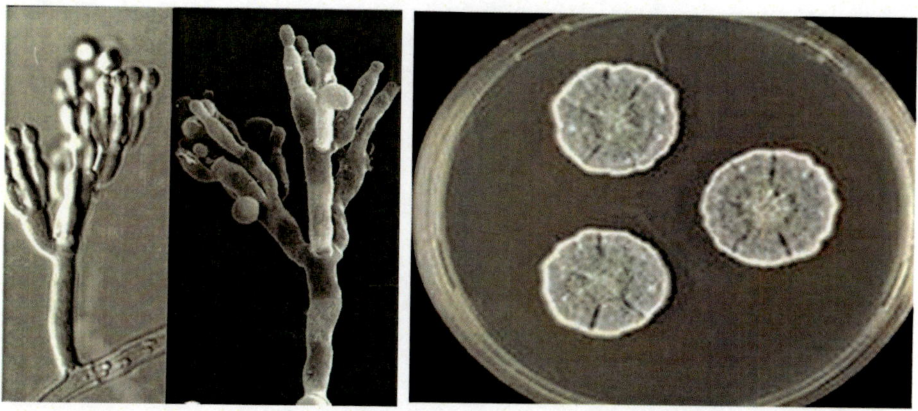

Fig. 257: Fungal colonies of *Penicillium griseofulvin* on media and structure of the fungus

This antibiotic is used in the treatment of dermatophytic diseases (diseases of skin, nails, hairs and feathers) like ringworms, athelete's foot and epidermophytics.

The compound was shown to be active against *Botrytis alli*, which causes neck rot in onions and it has been used for controlling *Botrytis* infection in agriculture. Different commercial products (Fig.258) are derived from this fungus.

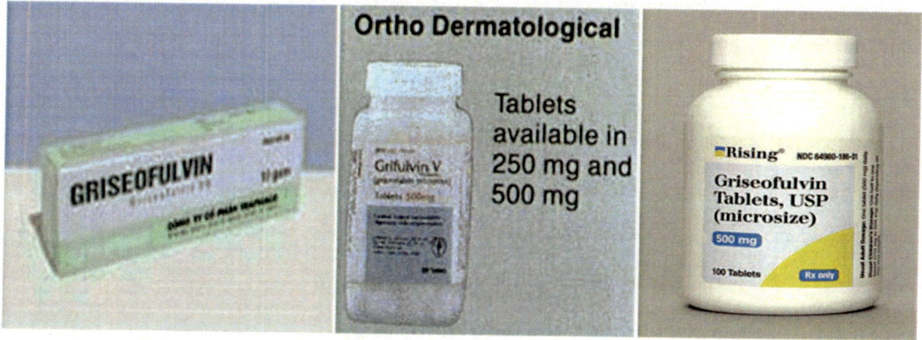

Fig. 258: Commercial Products derived from this fungi

12.3. *Acremonium* spp.

The fungus produces antibiotic **Cephalosporin.** The **cephalosporins** are a group of broad spectrum, semi-synthetic **beta-lactam** a class of antibiotics originally derived from the fungus *Acremonium* (Fig.259), which was previously known as "*Cephalosporium*".

This antibiotic is divided into three groups: Cephalosporin N and C are chemically related to **penicillins** and Cephalosporin P a steroid antibiotic resembles fusidic acid. Together with cephamycins, they constitute a subgroup of β-lactam antibiotics called cephems. Cephalosporins were discovered in 1945 and were first sold in 1964.

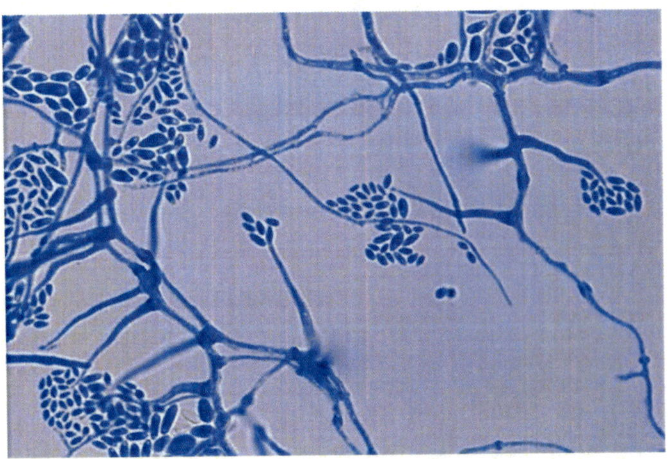

Fig. 259: Fungal structure and spores of *Acremonium* sp

Antibiotic Cephalosporins of various generations:

First Generation

Celospor, Celtol, Cristacef, Duricef, Ultracef, Keflex, Keftab, Kefglycin, Keflin, Cefadyl, Ancef, Kefzol, Velosef.

Second Generation

Ceclor, Ceclor CD, Distaclor, Keflor, Ranicor, Mandol, Monocid, Cefotan, Mefoxin, Cefzil, Ceftin, Kefurox, Zinacef, Zinnat.

Third Generation

Cefdiel, Spectracef, Claforan, Vantin, Cedax, Excede, Cefizox, Rocephin, Cefobid, Ceptaz, Fortum, Fortaz, Tazicef, Tazidime, Omnicef. Cefmax.

Fourth Generation

Maxipime, Cefrom

Fifth Generation

Zeftera, Teflaro

Combinations

Avycaz

Different commercial products (Fig. 260) derived from this fungus are used for treating different diseases.

First-line antibiotic treatment Ceftriaxone is used for the treatment of gonorrhoea, pelvic inflammatory disease and epididymo-orchitis. It is also an alternative to benzylpenicillin in patients with suspected meningitis.

Cefaclor may be considered as a second-line treatment for infections such as otitis media, sinusitis, cellulitis, diabetic foot infection and mastitis.

Cephalexin is a third-line treatment for urinary tract infection in pregnant women.

Fig. 260: Commercial products derived from the fungus *Acremonium*

Use of Cephalosporin in poultry and large animals

Poultry

Prevention of early chick mortality	20 gm daily for 3000 chicks for first 5 days in drinking water.
Treatment of Coryza, Fowl Typhoid, Fowl Cholera, Bacillary White Diarrhoea, E.Coli, Gangrenous Dermatitis and to combat secondary infection associated with viral diseases.	20 gm daily for 1500 chicks or 250 growers or 100 layers for 3-5 days in drinking water

Large Animals

Respiratory tract / Urinary tract infections eg: Pneumonia, Bronchitis, Cystitis, Nephritis, Pyrexia of unknown etiology etc	10 mg of Cephalexin per kg body weight 20 gm of Ceflax twice daily as a drench or with jaggery or with food.
Intrauterine Infections eg: Endometritis, Cervicitis, Pyometra, Retention of placenta, Repeat breeding etc.	20 gm to be dissolved in 50 ml of distilled water and given through Intrauterine route.

12.4. *Claviceps purpurea*

Claviceps purpurea fungi (Fig.261) produce Ergot alkaloids.

A large market is based mainly in Europe derives from ergot drugs; this market seems to be expanding as more and more pharmacological properties are discovered in the ergot drugs.

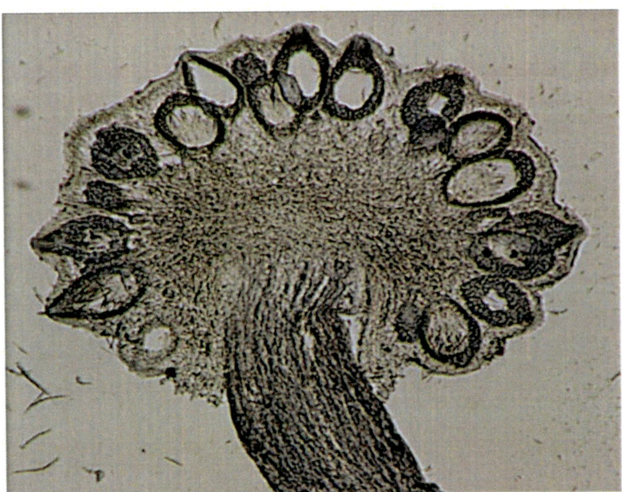

Fig. 261: Fungal structure of Claviceps purpurea

The alkaloid drugs derived from the fungi Claviceps purpurea are

Ergonovine

This is used to hasten labor pain and prevent postpartum bleeding. It is still used today and given to women in third stage of labor to prevent hemorrhaging by maintaining uterine contraction and tone. So they are used and misused to induce abortion.

Ergotamine

This compound produced by *Claviceps purpurea* is the powerful vasoconstrictor which means it constrict the blood vessels and thus the blood flow. This is useful in migraine headaches because it reduces extra cranial blood flow and the amplitude of pulsations in these arteries. They are used in throbbing headaches and cluster headaches.

Ergotomine is also used for relive for nausea and vomiting, Cancer treatments, to treat hepatitis B infection and to Treat high blood pressure

Different commercial products (Fig.262) are derived from this fungus.

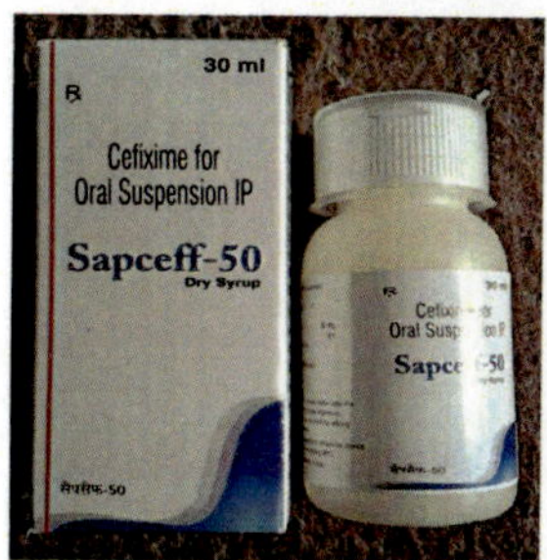

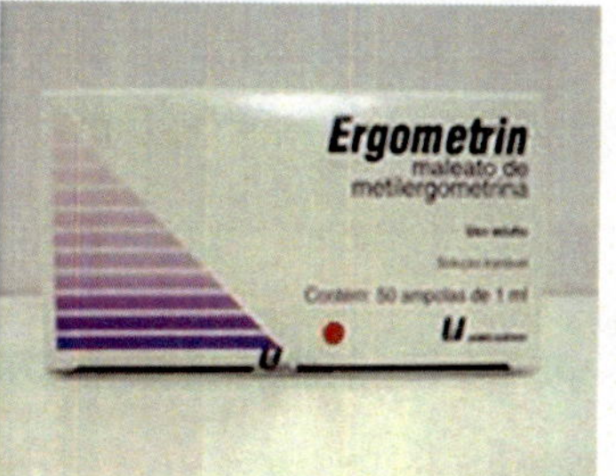

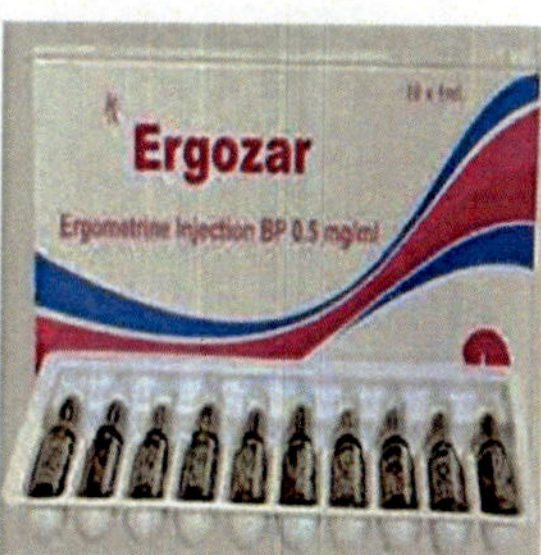

Fig. 262: Commercial products derived from the fungi Claviceps purpurea

12.5. *Aspergillus terreus* and *Penicillium citrinum*

A drug statin is produced from fungi *Aspergillus terreus* and *Penicillium citrinum*.

The generic names of the statin drug derived from these fungi are:

Generic Name of the drug	Brand Name of the drugs
atorvastatin	Lipitor
fluvastatin	Lescol
lovastatin	Altoprev, Mevacor
pitavastatin	Livalo
pravastatin	Pravachol
rosuvastatin	Crestor
simvastatin	Zocor

These drugs are used to Reduce levels of cholesterol in the blood.

12.6. *Tolypocladium inflatum*

The drug Cyclosporin is produced by the fungus Tolypocladium inflatum (Fig.263). Cyclosporin is immunosuppressant drug widely used in organ transplantation. It reduces the activity of immune system by interfering with the activity and growth of T cells.

Fig. 263: Fungal structure and spores of *Tolypocladium inflatum*

Brand Name of Drugs derived from the fungus *Tolypocladium inflatum*:

GRAFTIN SOFGELcap
IMUSPORIN cap
PANIMUN BIORALcap
PANIMUN BIORAL inj
SANDIMMUN NEORAL cap
SANDIMMUN NEORAL inj

These drugs are used as immune-suppressant which suppress or reduce the strength of immune system

Cyclosporin is used to prevent transplant rejection in people who have undergone kidney, heart and liver transplants.

It is approved in the US for the treatment of rheumatoid arthritis, sever psoriasis and as an ophthalmic emulsion for the treatment of dry eyes (restasis).

It is also used in sever atopic dermatitis, kimura's disease, pyoderma gangrenosum, chronic autoimmune urticaria.

Cyclosporin is highly effective in clearing psoriasis.

12.7. *Sacchromyces boulardii*

The fungus (Fig. 264) is used for the preparation of sacchromyces boulardii and related drugs (Fig.265).

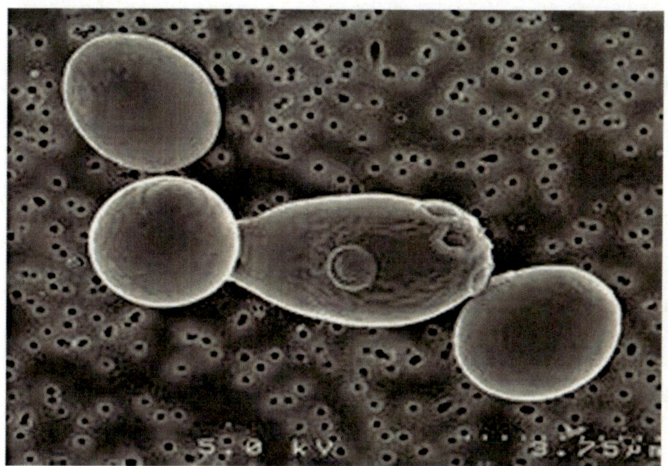

Fig. 264: Budding cells of *Sacchromyces boulardii* (in SEM)

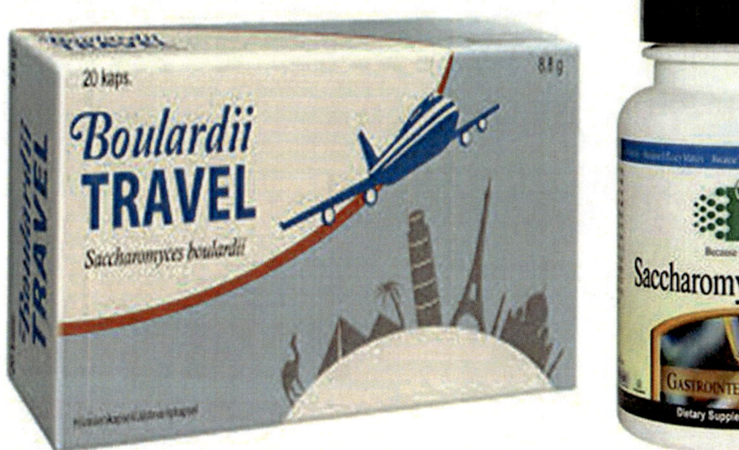

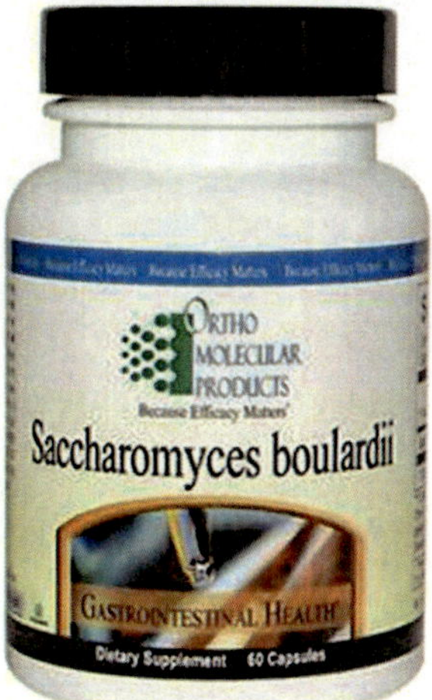

Fig. 265: Brand names of Drugs derived from the fungus *Sacchromyces boulardii*

These drugs are used against Acute diarrhea, Recurrent Clostridium difficile infection, Irritable bowel syndrome, Inflammatory bowel disease, Travelers' diarrhea, Antibiotic-associated diarrhea, HIV/AIDS-associated diarrhea and Elimination of Helicobacter pylori infection

12.8. *Coriolus versicolor* (Turkey Tails)

The fungus *Coriolus versicolor* ("multicolored mushroom"), (Fig.266) also known as *Trametes versicolor*, is a mushroom readily found in woodlands in China and Europe and is the most commonly found polypore in the oak woods of the Pacific Coast in the U.S. It grows in clusters or tiers on fallen hardwood trees and branches, frequently in large colonies. As its name implies, it is often multi-colored, with contrasting concentric bands, variously appearing in shades of white, gray, brown, black, blue or even red. It has a thin, velvety fruiting body, usually 2- 7 cm wide, fans out into wavy rosettes, giving rise to its popular name, Turkey Tails.

Fig. 266: Fruiting bodies of the fungus *Coriolus versicolor*

The product for pharmalogical use is known as PSK or PSP

The uses of PSK or PSP are

Anti-cancer action

PSK has been shown to be effective against several cancers, including cervical cancer, in combination with other therapeutic agents appears to enhance the effect of radiation therapy; PSP significantly lessened the side effects of conventional medical protocols used in the treatment of cancers of the esophagus, stomach and lungs, as well as significantly increasing the rate of remission in esophageal cancers.

Cardiovascular health

Lowered cholesterol in animal studies.

Immune enhancement

PSK increases interferon production, as well as scavenging superoxide and hydroxyl free radicals, has demonstrated anti-viral activity, possibly even inhibiting HIV infection.

12.9. *Ganoderma lucidum (Reishi)*

The fungus Ganoderma lucidum ("shining skin") (Fig.267) is a visually striking polypore with a hard woody texture and a shiny, varnished appearance. It primarily grows on oaks, plum trees and other hardwoods, and has a 2-20 cm semi-circular or kidney-shaped cap, variously colored white, yellow, blue, red, purple or black. Ganoderma species are found worldwide, though the Chinese and Japanese species have been studied the most extensively for their therapeutic value. It is somewhat rare in the wild, and so in recent years has been commercially cultivated, making it more widely available. In the West it is usually known by its Japanese name, reishi.

Fig. 267: Fruiting bodies of fungus *Ganoderma lucidam*

The uses of the fungus formulations are

Athletic performance

Enhances oxygenation of the blood, reducing and preventing altitude sickness in high altitude mountain climbers.

Cardiovascular health

Lowers cholesterol levels, reduced blood and plasma viscosity in a controlled study of patients with high blood pressure and high cholesterol.

Immune enhancement

Potent action against sarcoma, stimulates macrophages and increases levels of tumor-necrosis factor (TNF-α) and interleukins.

Immunopotentiation

Anti-HIV in in vitro and in vivo animal studies; protects against ionizing radiation.

Liver health

Reduced liver enzyme levels (SGOT and SGPT) in hepatitis B patients.

Respiratory health

60-90 % of 3,000 patients with chronic bronchitis showed clinical improvement, especially older patients with bronchial asthma. Regenerates bronchial epithelium (brochial tract lining).

12.10. *Polyporus umbellatus* (Zhu Ling)

The fungus *Polyporus umbellatus* ("umbrella-like polypore"), (Fig.268) also known as *Grifola umbellata*, is a white-to-gray mushroom that grows in dense rosettes from a single stem. It is found in deciduous woodlands in China, Europe and Eastern and Central North America, growing from dead tree stumps or the roots of birches, maples, beeches and willows.

Fig. 268: Fruiting bodies of the fungus *Polyporus umbellatus*

The commercial product is available as poly (Fig.269) which are used for different treatments

The uses of this fungus formulations are

Anti-cancer actions

Used in the treatment of lung and other cancers; has demonstrated pronounced anti-tumor activity in in vitro and in vivo animal studies; helps reduce the side-effects of chemotherapy.

Immune enhancement

It stimulates and enhances the performance of the immune system and accelerates production of IgM and strengthens the power of monocytes.

Liver health

It can help alleviate symptoms of chronic hepatitis; was used as part of an herbal formula that cured 17 of 39 patients with cirrhosis of the liver, and brought about significant improvement in 19 others.

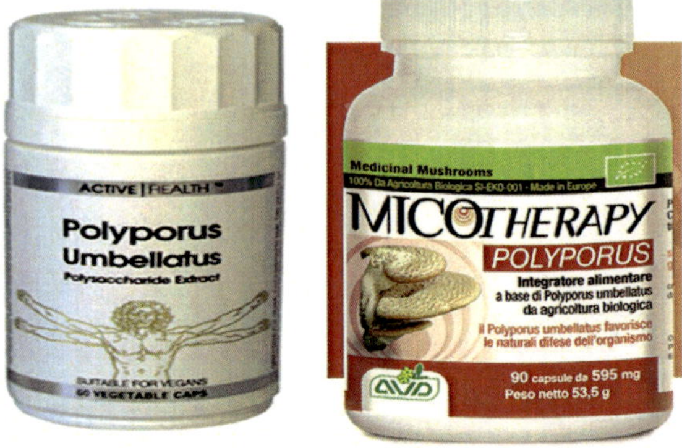

Fig. 269: Pharmacological product derived from the fungus polyporus umbellatus

12.11. *Hericium erinaceus* (Lion's Mane)

Hericium erinaceus ("spiny hedgehog") is a snow-white, globe-shaped fungus composed of downward cascading, icicle-shaped spines. Its striking appearance gives rise to its various common names, Lion's Mane (Fig.270), Monkey's Head and Hedgehog Fungus. It grows up to 40 cm in diameter on dead or dying broadleaf trees such as oak, walnut, maple and sycamore and is found in China and Japan, as well as parts of Europe and North America. It is considered a gourmet mushroom, long popular with forest folk, with a flavor variously described as reminiscent of lobster or egg plant.

Fig. 270: Fruiting body of the fungus *Hericium erinaceus*

The uses of this fungus formulations are

Anti-cancer effects

Helps in the treatment of esophageal and gastric cancers, may extend the life-span of cancer patients.

Digestive enhancement

Promotes proper digestion; effective against gastric and duodenal ulcers and gastritis.

Immune enhancement

Protects the gastrointestinal tract against environmental toxins, inflammation and tumor formation, an extract was used as part of a protocol that helped increase T and B lymphocytes in mice.

Culinary use

Hericium erinaceus is a choice edible when young, and the texture of the cooked mushroom is often compared to seafood. It often appears in Chinese vegetarian cuisine to replace pork or lamb. This mushroom is cultivated commercially on logs or sterilized sawdust and is available fresh or dried in Asian grocery stores.

Medical research and use

Hericium erinaceus has a long history of use in traditional Chinese medicine. A 2005 rat study showed some compounds in the mushroom, like threitol, D-arabinitol, and palmitic acid, may have antioxidant effects, regulate blood lipid levels and reduce blood glucose levels. A 2012 study on rats that had suffered brain injury showed that «daily oral administration of H. erinaceus could promote the regeneration of injured rat peroneal nerve in the early stage of recovery. More recently and more relevant to human use, is a 2013 review of scientific studies, which asserted the medical benefits of the mushroom by saying "This mushroom is rich in some physiologically important components, especially β-glucan polysaccharides, which are responsible for anti-cancer, immuno-modulating, hypolipidemic, antioxidant and neuro-protective activities. H. erinaceus has also been reported to have anti-microbial, anti-hypertensive, anti-diabetic, wound healing properties among other therapeutic potentials." A 2014 scientific review on the therapeutic effects of H. erinaceus concluded that "it is helpful to various diseases, such as Alzheimer's disease, immunoregulatory, and many types of cancer."

A report reveals that pills of this mushroom are used in the treatment of gastric ulcers and esophageal carcinoma. A 2011 study on rats demonstrates the mushroom's wound healing capacities. Considering the increase of degenerative conditions, scientists around the world have launched investigations on the possible anti-dementia compounds of this mushroom.

12.12. *Aspergillus versicolor*

Aspergillus versicolor fungi (Fig.271) produce antibiotic Versicolin.

The antibiotic is specifically active against pathogenic fungi *Trichopyton ruburum* which cause 90 per cent of skin infections occurring in Calcutta and eastern India.

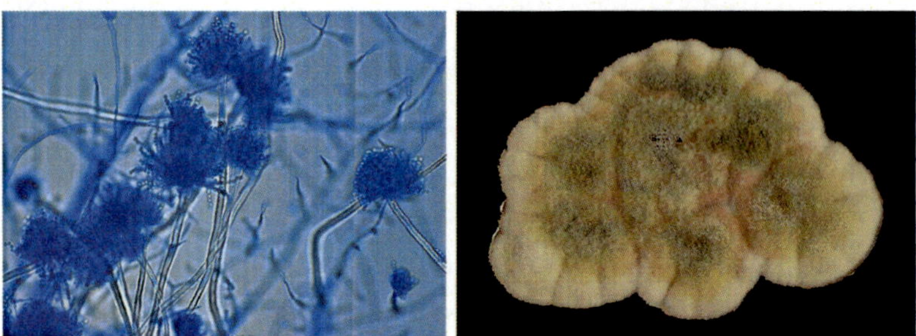

Fig. 271: Fungal structure of *Aspergillus versicolor* and growth on media

12.13. *Oudemansiella mucida* and *Stobilurus tenacellus* (Fig.272)

oudemansiella mucida fungi produce antibiotic Strobilurin A and *Stobilurus tenacellus* fungi produce antibiotic Strobilurin B.

The anti- fungal activity, especially of stobilurins B in vitro and in the greenhouse against plant pathogens like *Venturia inaequalis, Cercospora arachidicola, Plasmospara viticola and Phythophthora infestans* was excellent.

Fig.272: *Fruiting bodies of the fungus Oudemansiella mucida* and *Stobilurus tenacellus*

12.14. *Trichoderma virens*

Trichoderma virens fungi (Fig.273) produce antibiotic Gliotoxin.

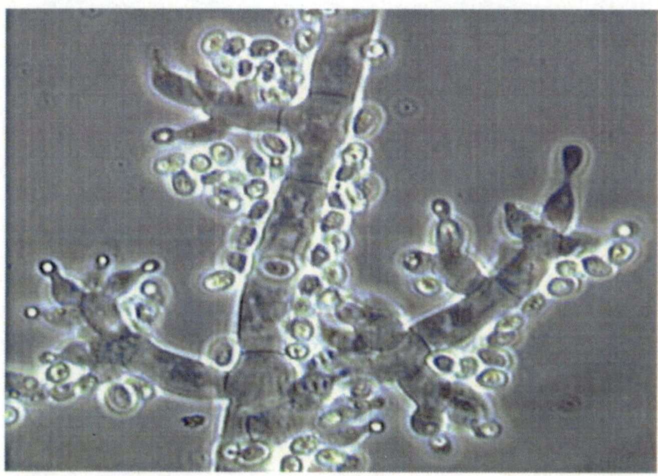

Fig. 273: Fungal structure of *Trichoderma viren*

The antibiotic is used to control plant pathogenic fungal infection of *Rhizoctonia solani*.